Lecture Notes in Social Networks

Lecture Notes in Social Networks (LNSN) comprises volumes covering the theory, foundations and applications of the new emerging multidisciplinary field of social networks analysis and mining. LNSN publishes peer- reviewed works (including monographs, edited works) in the analytical, technical as well as the organizational side of social computing, social networks, network sciences, graph theory, sociology, semantic web, web applications and analytics, information networks, theoretical physics, modeling, security, crisis and risk management, and other related disciplines. The volumes are guest-edited by experts in a specific domain. This series is indexed by DBLP. Springer and the Series Editors welcome book ideas from authors. Potential authors who wish to submit a book proposal should contact Annelies Kersbergen, Publishing Editor, Springer

e-mail: annelies.kersbergen@springer.com

Sibel Tarıyan Özyer • Buket Kaya
Editors

Cyber Security and Social Media Applications

Springer

Editors
Sibel Tarıyan Özyer
Department of Computer Engineering
Ankara Medipol University
Ankara, Türkiye

Buket Kaya
Department of Electronics and Automation
Fırat University
Elazig, Türkiye

ISSN 2190-5428 ISSN 2190-5436 (electronic)
Lecture Notes in Social Networks
ISBN 978-3-031-33064-3 ISBN 978-3-031-33065-0 (eBook)
https://doi.org/10.1007/978-3-031-33065-0

This Springer imprint is published by the registered company Springer Nature Switzerland AG
The registered company address is: Gewerbestrasse 11, 6330 Cham, Switzerland

Contents

Powering Blogosphere Analytics with BlogTracker: COVID-19 Case Study

Abiola Akinnubi, Nitin Agarwal, Mainuddin Shaik, Vanessa Okeke, and Ayokunle Sunmola

Abstract Tracking and analyzing large dataset from various blog collections has become a common challenge for analysts, social scientists, information scientists, behavioral scientists, industries, and government agencies. Since Blogging has expanded exponentially, various platforms now incorporate blogging to extend their websites and applications. Blogging has since become where people share thoughts and seek to influence and push narratives to diverse sections of the public, customers, followers, and others. The blogosphere is a virtual community of blogs that exists together with influence spanning beyond the borders of countries and regions and is not restrictive in the length of what bloggers can post; this allows various authors to generate content since blogs provide an avenue to shape narratives and influence due to their unlimited capabilities and are free from government regulation and controls. Blogs create a treasure trove of information that can be analyzed and generate meaningful insights through various visualization mechanisms and information science approaches.

Keywords BlogTracker · Social media · COVID-19 · Vaccines · Blogosphere · Narrative analysis

1 Introduction

The blogosphere has been a virtual community with deliberation and conversation around various topics of interest since narratives, influence, and sentiments around hot topics have migrated to online communities from the traditional media and physical interactions. Blogging has given birth to the rise of virtual engagement in the blogosphere by various types of users, i.e., bloggers and commenters or

A. Akinnubi (✉) · N. Agarwal · M. Shaik · V. Okeke · A. Sunmola
Collaboratorium for Social Media and Online Behavioral Studies, University of Arkansas, Little Rock, AR, USA
e-mail: asakinnubi@ualr.edu; nxagarwal@ualr.edu; vokeke@ualr.edu; seayokunle@ualr.edu; mxshaik@ualr.edu

S. T. Özyer, B. Kaya (eds.), *Cyber Security and Social Media Applications*,
Lecture Notes in Social Networks, https://doi.org/10.1007/978-3-031-33065-0_1

followers since blogs are free from the long historical censorship and regulation of traditional media in many parts of the world. Kolari and Finin [10] described blogs to composed of dated submissions typically organized in transpose chronological bid on a single web page. Kolari and Finin [10] defines blogs as tall influence and specialized releasing infrastructure the subset of the blogosphere.

Traditional public media generally pushes the narratives of the primary stakeholders, such as governments and private individuals who founded their media or funds their media for their own political and socio-cultural beliefs. This common practice pushed the public to adopt alternate platforms. Blogs are one of such alternate platforms where bloggers, content, and followership have been the driving force of the web, and followership has been the driving force of the web-based virtual community. Singh et al. [15] described the challenge with most publicly accessible tools for most blog analytical which focuses more on marketing applications. With efforts on an encompassing exploitation of the blogosphere are nearly negligent. According to Singh et al. [15], since blogs are unique free-form writings, they are filled with extremely modern and emotion laden; composes the blogosphere an ideal forum for socio-political analysis.

Blogging allows authors to share an infinite amount of information and content, be it a video, texts, or images, thereby becoming a powerful tool in forming narratives and influencing audiences who share similar beliefs and societal values. Analysts have had to put together many makeshift tools or third-party libraries, codes, and software to study the infinite amount of daily data generated in the blogosphere. There is difficulty finding free publicly available resources with the all-in-one capability of following events and activities in the blogosphere or analyzing various discussions, extracting narratives, visualizing narratives, detecting keywords of hot topics, highlighting influential bloggers, and understanding sentiments that each blog post and blogger push. These challenges motivated us to develop the BlogTracker Application to cover various aspects of online behavioral studies and with the capability to reduce the workload that analysts traditionally go through in understanding what is happening around the world. In Akinnubi et al. [2], the authors demonstrated some capabilities of BlogTracker, but the research focused mainly on the influential campaigns without expanding the scope to study various aspects of the blogosphere.

This work expands our earlier work Akinnubi et al. [2], by demonstrating how the BlogTracker Application can power the blogosphere analysis. We use COVID-19 as a use case to demonstrate BlogTracker's capabilities by explicitly showing how COVID-19 narratives were formed—and visualized among other critical salient features that BlogTracker can perform. The BlogTracker application allows analysts to focus on more important things than hacking various makeshift solutions. Compared to other tools that have either gone into extinction or do not offer public accessibility freely or without restricting their usage unless the user purchases a paid license, BlogTracker has emerged to solve these various challenges and many more.

2 Literature Review

Over the years, there has been an increase in the blogosphere studies, but new insights emerge daily in our ever-changing world with more studies. Bloggers and audiences provide opinions and reactions to new information, where studies available on various aspects of the blogosphere like narratives, topic modeling, content diversity, blog posting frequency, and sentiment analysis. Mining a significant scale of opinions from blogs has its challenges, especially with the scale at which blog posts are churned out daily by bloggers due to its unregulated nature and the bloggers being the traditional regulator of their blog sites themselves.

The concept of text-mining is key to how blogosphere is studied. Since, collecting data and tracking blogs fits into a sub-field of text-mining application specifically referred to as blog mining according to (Hassani et al. [7]). Hassani et al. [7] also highlighted existing challenges in continuous computation analysis on new stream of data compared to traditional computation and analysis run on static collected blog data. Their work highlighted the need for real-time solution for analyzing blog data. Singh et al. [15] described the limitation in mechanism of analyzing blog data by describing existing tools as limited to syntactic mechanisms. Their work also highlighted the lack of sophisticated semantic tools capable of performing natural language processing (NLP) task. With the availability of such describe to make analyzing much more valuable.

Tsirakis et al. [16] developed a tool for social, news, and blog data but emphasized business, where the authors proposed a tool called "PaloPro" that focuses on the platform as a service for business or brand monitoring. Cuong et al. [12] explores a learning analytic tool to study blogosphere using two approaches: blogosphere structural analysis and content analysis, with the help of a tool called Mediabase and eTwinning Network by applying social network analysis methods leveraging these tools in studying the blogosphere while using the Mediabase for content analysis. Shaik et al. [13] explored the Australian blogosphere using a multi-method analytical framework to cover various aspects of the Australian blogosphere. Shaik et al. [13] analyzed over 20,066 blog posts and 10,113 comments between 2019 and 2020 and found that COVID-19 discourse absorbed much of the blogger's attention during the study period.

Agarwal et al. studied the Influential bloggers in a virtual community by aggregating bloggers having similar interests, where the authors treated these bloggers as a virtual community in the blogosphere and observed that similar and influential bloggers' communities clustered. Bansal [5] Also discussed is the concept of long-tail on the relationship of the influential blog site. He concludes that influential blog sites are few, which depends upon the nature of the blogosphere.

A study by Hussain et al. [8] on analyzing the shift in narrative explored the concept of narratives with a focus on migrants to Europe; this allowed the authors to use targeted sentiment analysis to study how narratives have shifted toward migrants in the European blogosphere with over 9000 blog posts studied from 2005 to 2016. Hussain et al. had used the migrant crisis in the European union to understand how narratives were weaponized by extensively monitoring citizens' sentiment change towards migrants, and they understood how the shift of narratives occurred in the European union blogosphere.

Al-Ani et al. [3] explored the concept of counter-narratives in the Egyptian blogosphere during the Arab uprising by using topic modeling, where the study explored and looked at blogs across the interaction of societal, personal, and revolutionary blog topics from 2005 to 2011. Their works found that bloggers could organize counter-narratives against governments' narrative push. The work of Al-Ani et al. [3] was able to explore how blogs could be a suitable mechanism in counter-power wielding since blogs are neither dependent nor regulated by authorities like the mainstream media since blogs are not susceptible to traditional gatekeepers and character limitations of micro-blogging platforms.

Bandeli et al. [4] proposed a framework using NLP techniques in identifying actors and actions and POS (Parts of Speech Tagging) tagging since social media users can create narratives and spread information using blogs. The framework allows for effective counter-narratives to reduce the impact and spread of propaganda campaigns. Although there were some limitations with some grammar rules that needed to be updated if the sentence was complex, their work demonstrated the narrative extraction from entities such as person, organization, and location constructed in a sentence and events in the blogosphere. This work extends the narrative extraction approach by allowing users edit and merge narratives through visualization mechanisms allowing an extra layer of quality control when a complex statement is encountered.

3 Methodology

To study the blogosphere, we considered available tools that provide analyst the capabilities to leverage solutions that were developed from years of research and provided better visualization with no cost. At the same time, there were either underdeveloped, non-frequent updated, non-robust or expensive paid tools. We propose a BlogTracker tool for end-to-end social data analysis of the blogosphere, where it allows users to follow blogs of interest and track various hot social discussions. In addition, BlogTracker allows users to perform various analytic and insight mining from blog data that the user chooses to track or follow using the information or keyword the user intended to track. It is a state-of-the-art tool capable of providing results in areas not limited to cluster analysis, sentiment analysis, narrative analysis, topic modeling, keyword trend, and influence analysis, amongst other features, since the blogosphere is an entire virtual universe of its own shaped by community participatory of bloggers aka Authors, commenters and in some situations by policymakers. Then mandates studying various topics discussed within this virtual global community and how they connect to what is happening worldwide since the blog provides influencers and opinion-makers with the ability to shape narratives with the volume of content, and any local or international laws do not restrict it. We have developed several modules in our BlogTracker application that analysts can leverage to analyze what is going on in the blogosphere's virtual world. In showing the capabilities of the BlogTracker application, we will discuss and demonstrate how various features of the BlogTracker application helps analysts understand the entire blogosphere applications.

3.1 BlogTracker

Our BlogTracker web app is developed based on our capability to ingest extensive data sets from blog sites worldwide with a resilient data pipeline in the background to handle data traffic and structure this data properly into a temporary data store. The data is then ingested by a threaded data post- processing and preprocessing layer capable of crunching a massive dataset in a matter of minutes that is developed on Java. Due to the nature in which these data sets are retrieved, we layered our application on both MySQL database and NoSQL document database to serve the elastic search layer for faster data fetch and retrieval. Our backend shown in Figs. 1 and 2 seats Java and spring boot which provides a structured way of architecting an application capable of crunching large dataset retrieved from the blogosphere. We use Nodejs and expressjs serving as our frontend architecture for rending data visualization and analysis for the users' Figs. 1 and 2 shows a high-level overview of our technical architecture and how the BlogTracker application is served to the end-user.

3.1.1 Narrative Analysis

To extract and analyze narratives from blog data in the Tracker of study, we adopt our framework for narrative analysis as explained in the work of Hussain et al. [9]. Hussain et al. [9] and shown in Fig. 3. This feature allows users to see the top ten keyword entities extracted along with each associated narrative and can then be viewed along with the associated blog post. The tool also enhances the user's capability by searching for keywords associated with a tracker. Search returns the keyword match associated with the user search criteria, enabling users to visualize narratives.

3.1.2 Influence Analysis

The influence analysis feature attempts to find bloggers and topics of influence. The influence analysis feature examines how blogs and bloggers propagate information, use their influence, score their influence, and visualize their influence in the blogosphere. This feature Fig. 4 helps identify a blogger or blog post's influence on the blogosphere.

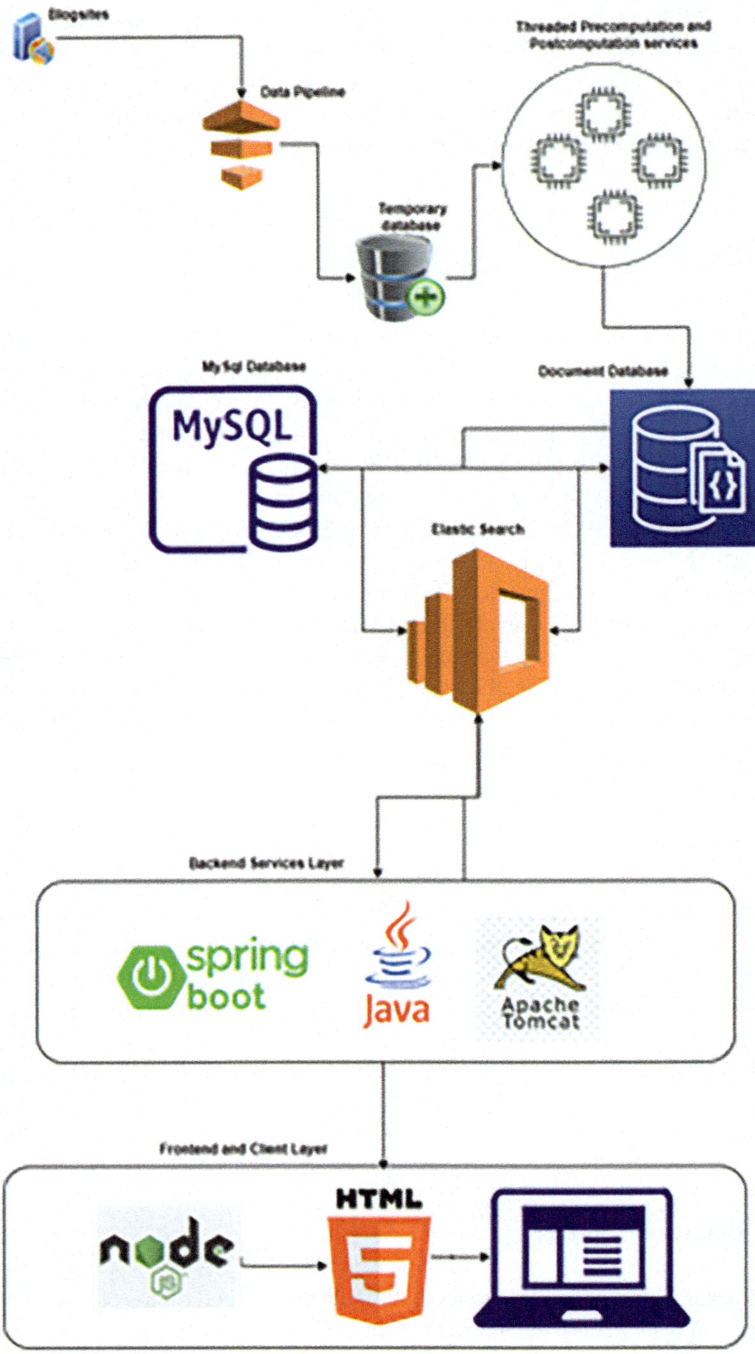

Fig. 1 High-level architecture overview of the BlogTracker application

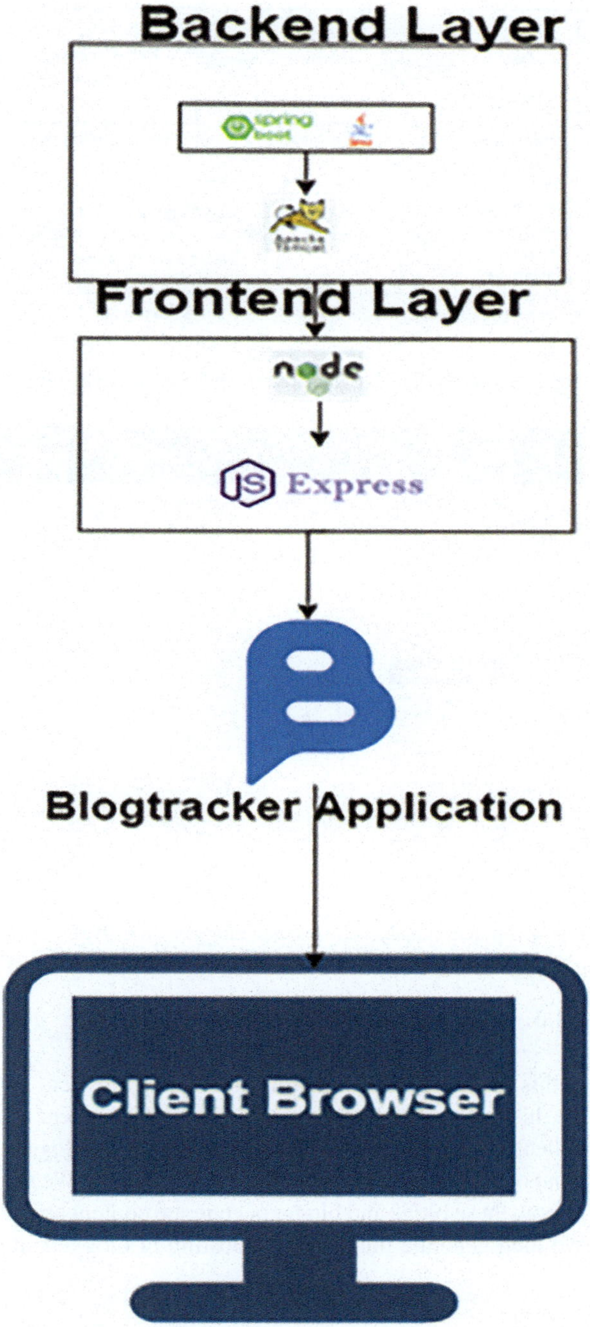

Fig. 2 An high-level architecture overview of the BlogTracker application

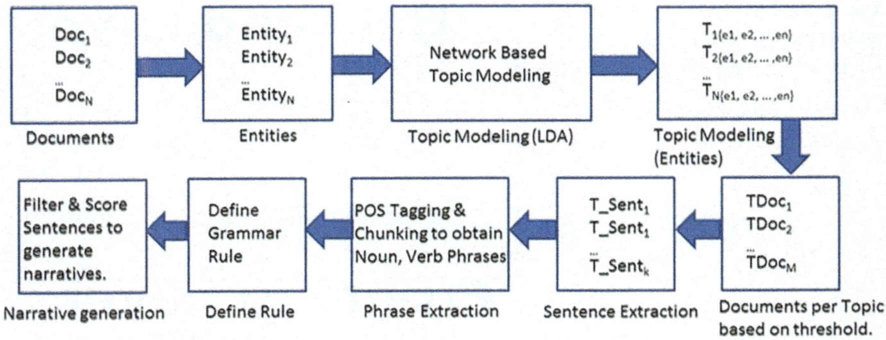

Fig. 3 Framework to extract narratives adapted after Hussain et al. [9]

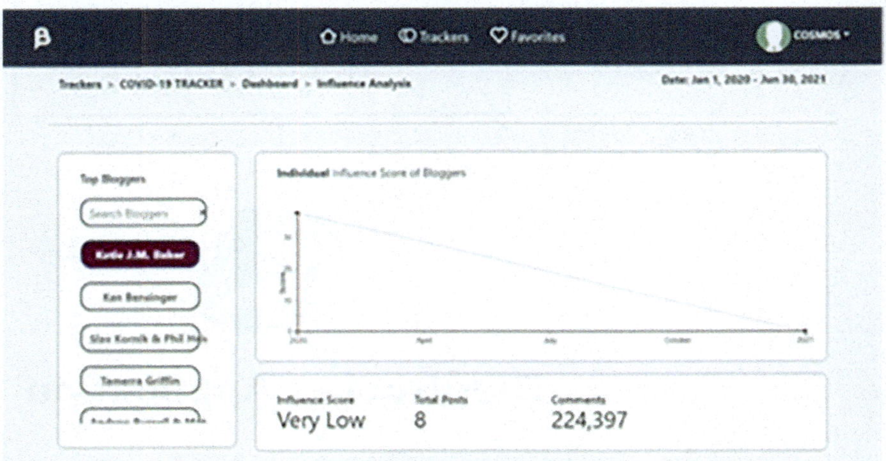

Fig. 4 Influence analysis of selected bloggers on COVID-19 tracker

3.1.3 Blogger and Blog Portfolio

In this approach, for generating and having bloggers' portfolios, we aimed to decode the history and patterns that blogs, and bloggers could have to shape the data collected around the COVID-19 dataset in the blogosphere. We sought to generate what overall sentiment blogs have and bloggers. We also create a weekly breakdown of how these blogs and bloggers generate content around interest in the blogosphere. We then generate the historical posting of bloggers and blogs and provide insights into questions such as "What is the posting pattern?", "Do blogs or bloggers post more at the beginning, middle, or end of the year?", "How often do blogs and bloggers post?" Figures 5 and 6 show blogs and bloggers that posted more during the cause of the coronavirus pandemic. Blogger and Blog Portfolio can study Daily, Monthly, and Yearly patterns to derive more accurate assumptions from blog posts.

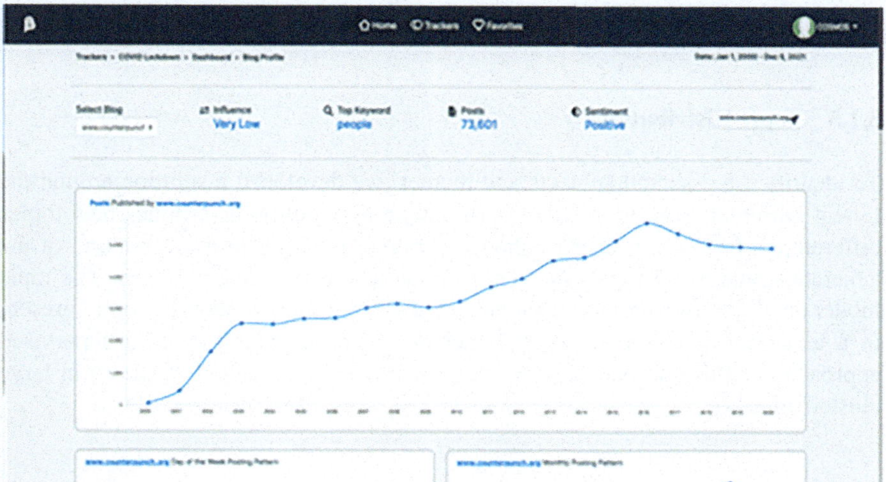

Fig. 5 BlogTracker's blog portfolio

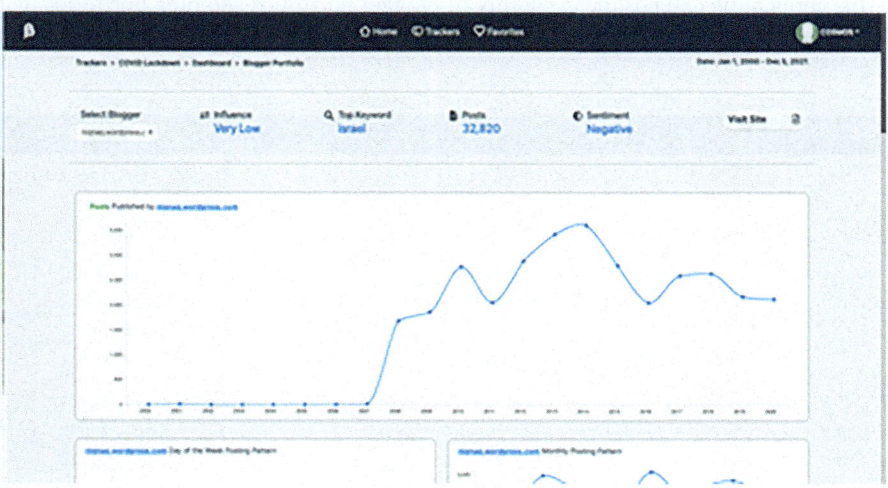

Fig. 6 BlogTracker's blogger portfolio

3.1.4 Sentiments Analysis and Emotion Analysis

We used our sentiment and emotion analysis feature during this study, built on various social and behavioral Likert scales. This analytical tool helps display the trend of sentiments categorized into positive and negative sentiments with different emotion ratings like joy, sadness, trust, and anticipation on over 383 blogs for the selected period in the COVID-19 Tracker. Figure 7 shows the dashboard obtained from our BlogTracker tool.

3.1.5 Topic Distribution

To identify the essential subjects and themes, we developed a solution around the Latent Dirichlet Allocation (LDA). This topic distribution determines how topics differentiate from each other as shown in Fig. 8. We also limit our ranking of the generated categorical topics to 10 topics due to massive data volumes. The topic model involves discovering hidden patterns in text corpora to identify topics present in a text object, which assists in better decision making. It is an unsupervised approach used to find and observe many words (usually called "topics") in large clusters of texts.

3.1.6 Content Diversity Analysis

This application extends the capability of blogger portfolios and blog portfolios that use the computed LDA topic distribution data to compute for novelty, transience,

Fig. 7 BlogTracker's sentiments analysis and emotion analysis

Fig. 8 BlogTracker's topic distribution

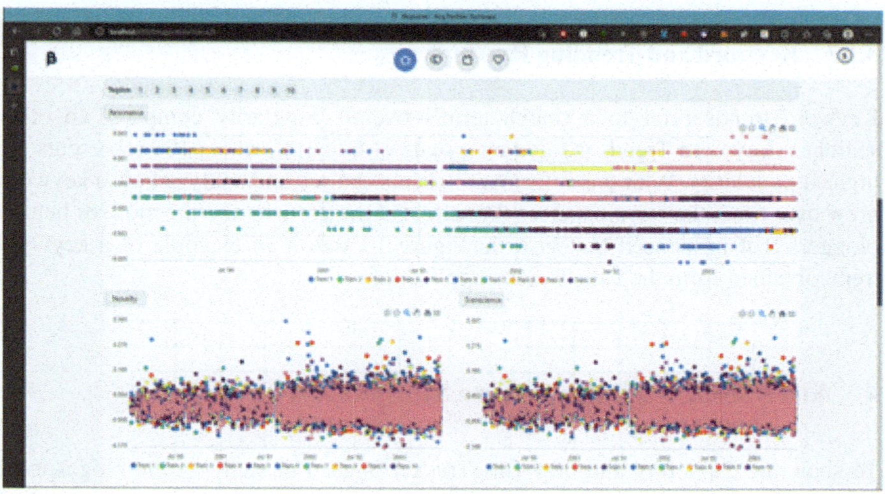

Fig. 9 Image of content diversity analysis capability of the BlogTracker application

and resonance, which was first defined in Barron et al. [13] and the works of Stine and Agarwal [14]. This analysis helps analysts identify shifting discursive priorities within a blog over time. Figure 9 shows the visualization within the BlogTracker application to understand the blogosphere shifting discursive.

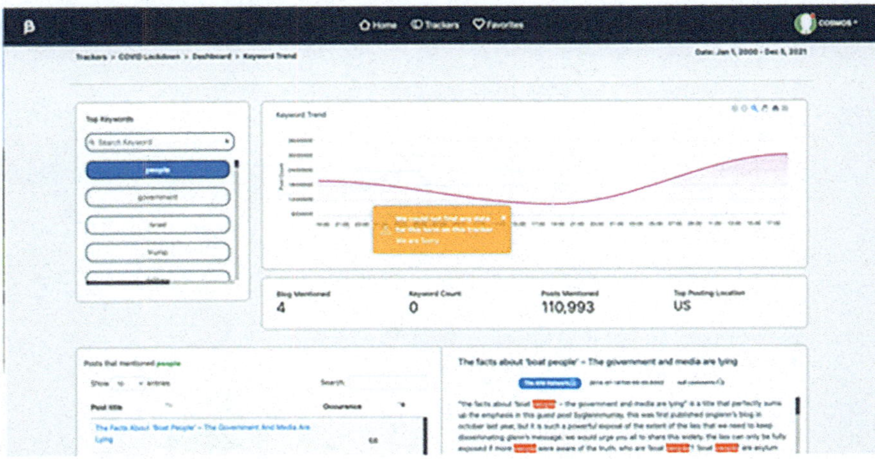

Fig. 10 BlogTracker's keyword trend

3.1.7 Keyword and Trending Hot Topics

Keyword trends refer to a search term's overall popularity compared to other searches. Keyword Trends is rated on a scale of 0 to 100, where 100 represents the highest popularity. With this information, users can see and analyze how a keyword grew over time, the circumstances that affected the growth, and the motives behind bloggers that influenced that growth. Figure 10 shows an example of a keyword trend obtained from the BlogTracker application.

4 Analysis and Findings Case Study: COVID-19

To show the capability and how BlogTracker is used in studying the blogosphere to track the COVID-19 discourse, we look at a tracker that contains a total of 383 unique blog sites and a total of 852,106 blog posts; the number of bloggers contained in this specific Tracker is 64,520. Table 1 shows this tracker information, while Fig. 11 shows how this information is presented to analysts in the BlogTracker application. Figure 12 shows the dashboard summarized information of the entire blogosphere of the selected Tracker with 383 unique blog sites we were able to

Table 1 Key statistics of the COVID-19 tracker used in the study

COVID-19 tracker	Statistics
Blog sites	383
Blog post	852,106
Number of bloggers	64,520
Number of comments	11,392,673

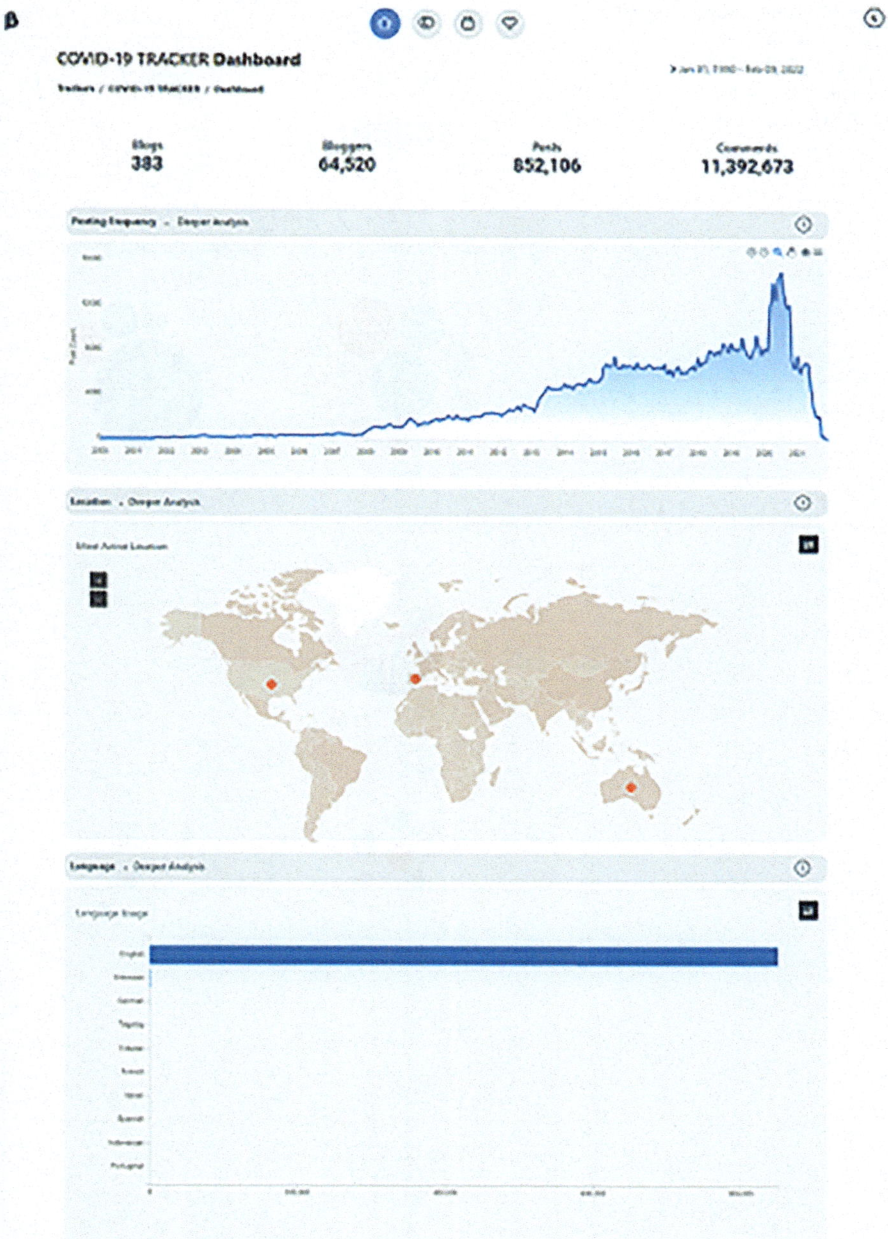

Fig. 11 Shows dashboard statistics at the top with summary statistics of information of COVID-19 tracker

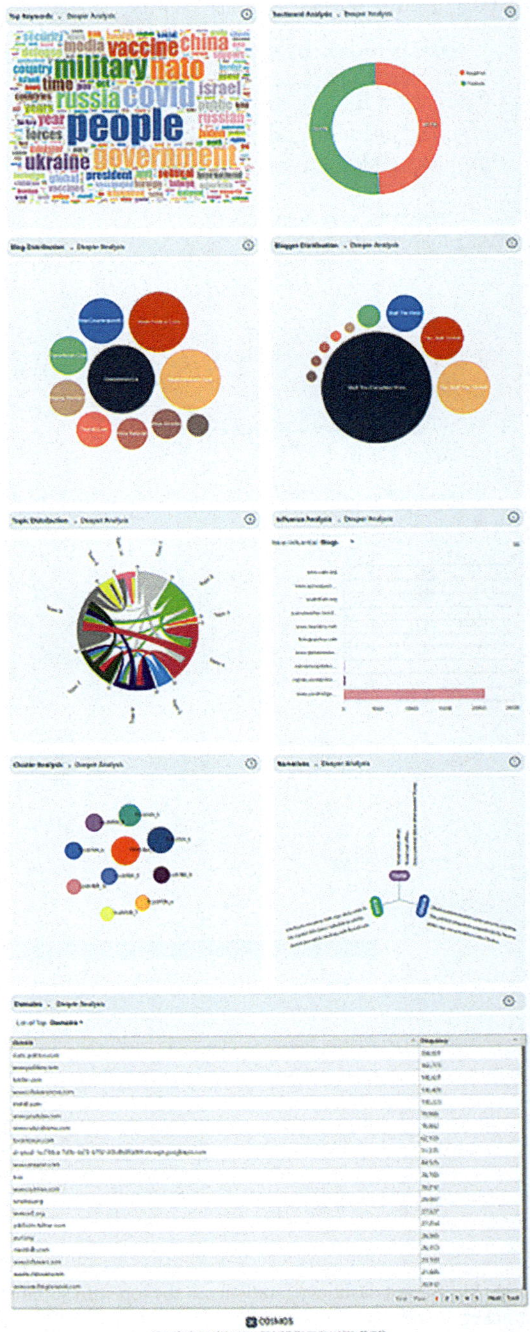

Fig. 12 BlogTracker dashboard showing the concise yet detailed visualization of the COVID-19 tracker

Tracker under this Tracker, with the information presented in a way that gives analysts a concise view but detailed of the hot discussion under a topic that was discussed during the COVID-19 pandemic. The dashboard provides other essential information like the top posting location where the topics were discussed, the top language used in generating posts,top narratives and their extracted entities, top bloggers, and much other essential information.

4.1 Cluster Analysis

We used K-mean clustering from the studied blog site data and selected only the top 10 clusters for visualization with Fig. 13, showing cluster 1 as the dominant cluster and statistics associated with cluster 1 selected. As we can see from the first cluster visualization in Fig. 13, the top posting location for the blogs in this cluster is the United States, and it has a total of 275 bloggers mentioned in over 128,000 blogs; these statistics are data obtained over the time of February 8, 2022, till February 11, 2022. The U.S. is the top posting location where bloggers have contributed the most during the study period. While the posting location may be where the blog site server is hosted, we use the location captured as the posting location that has contributed the most to the topics discussed in cluster 1. We also presented the top keywords that shape the various discussions in cluster 1 in the word cloud and how these common terms overlapped in Fig. 14. During the pandemic, words like health, people, Canada, public, American, governments, presidents, and many more were the center of attraction as the dominant keywords in cluster 1. We also show some of the top blog posts in Fig. 15 for analysts to visualize the blog post from each cluster and correlate this with the various keywords shown in Fig. 13. An analyst can also view the blog post content itself, and This table gives researchers the flexibility to view all blog posts associated with a selected cluster and do further analysis using other tools on the BlogTracker application.

4.2 Posting Frequency

We also track and analyze the posting pattern of various blog sites and authors using our feature called posting frequency. With the posting frequency Fig. 16, we observed the globalnews.ca as one of the top contributors with a peak posting trend of globalnews.ca peaking between 2013 to 2020 before a steady decline afterward, demonstrating that this blog site seems to have lesser contribution around COVID-19 after the global pandemic slowed down. We also compared the posting trend performance of globalnews.ca to other top blogging sites and noticed similar behavior, although globalnews.ca's peak posting period outperformed other

Fig. 13 Showing 10 clusters and keywords in the word cloud and common terms chord overlaps in various clusters

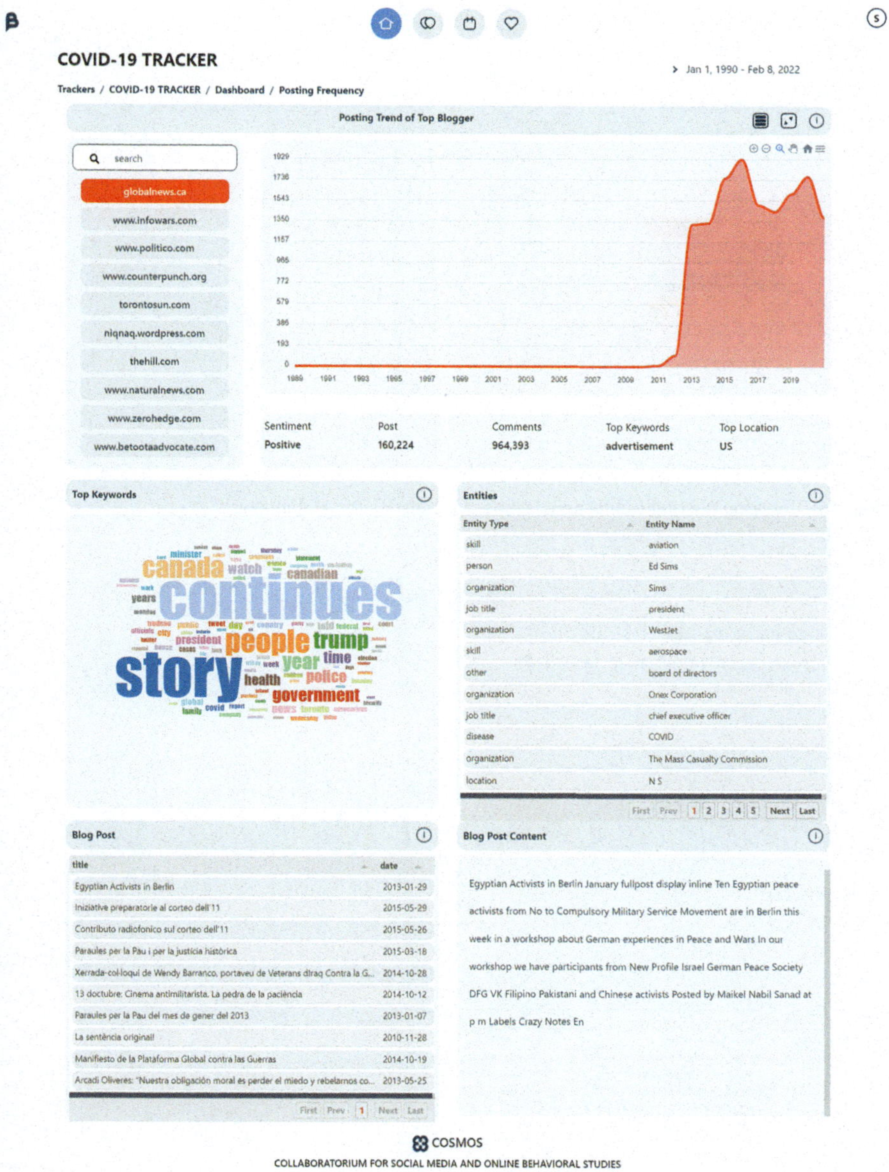

Fig. 14 Showing comparison of various top blog site posting trends

top sites like inforwars.com and counterpunch.org. However, when backtracking the visualization in Fig. 17, which shows these comparisons, we noticed that counterpunch.org seems to have been an early blogger around pandemic and health-related discussions. From the statistics tab of the selected blog post below, we see that the overall sentiment is positive for globalnews.ca; included in the statistics is

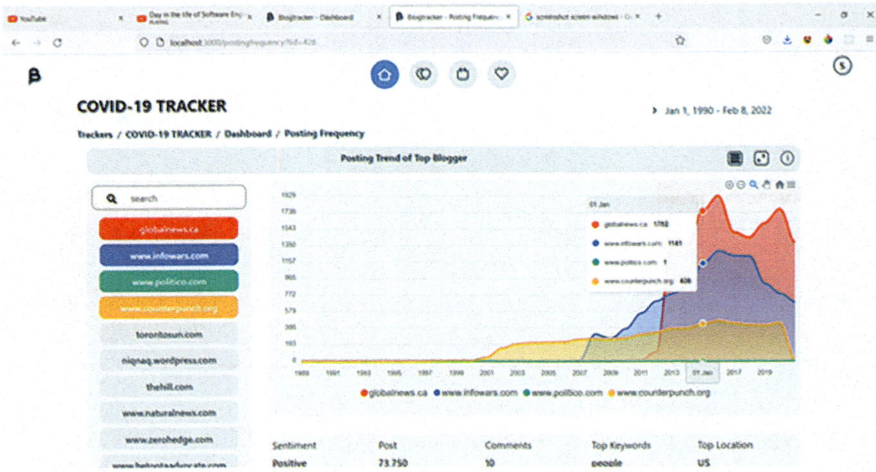

Fig. 15 Showing posting trends and dashboard statistics for globalnews

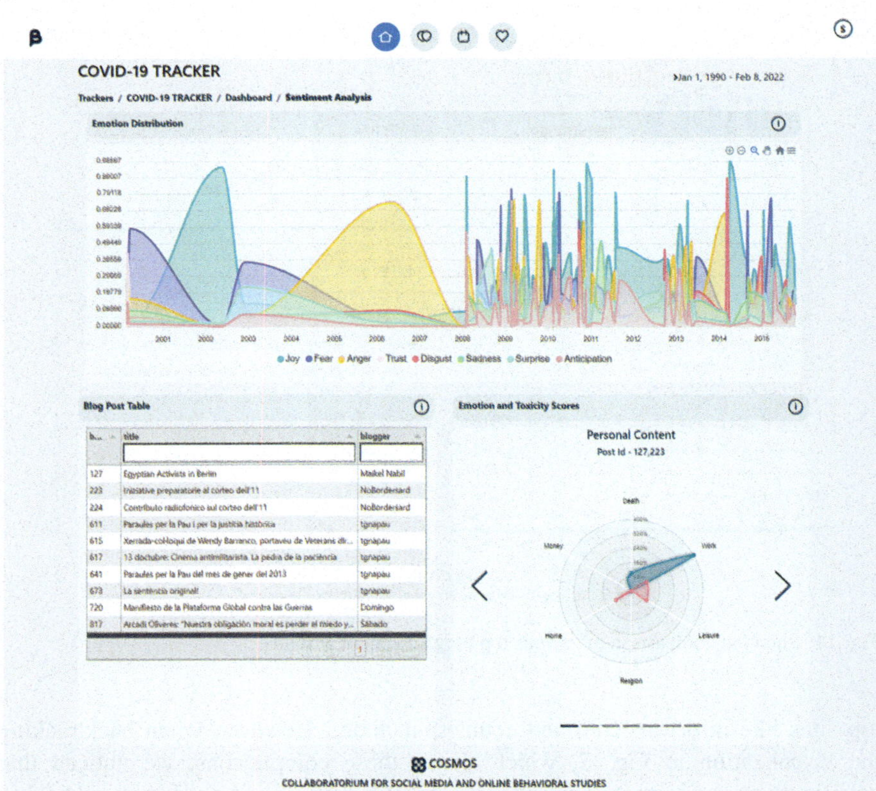

Fig. 16 Emotion distribution visualization using the Likert scale emotions variables for rating reactions

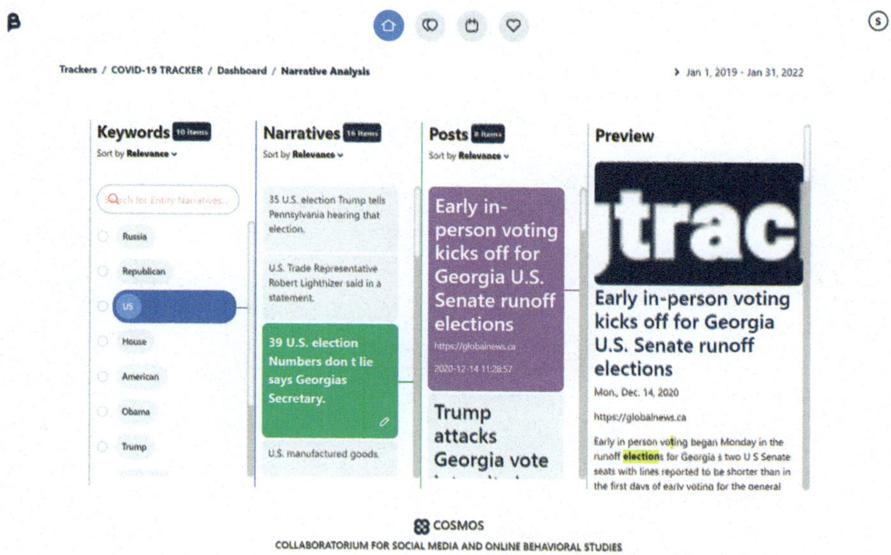

Fig. 17 Narrative visualization and highlighted narratives with the blog post content

the number of posts on the selected blog site, 106,224, the number of comments on the blog site, 964,393 and the top keyword from this unique blog site is seen to be ‘advertisement’. The top keywords for this posting trend of the selected Tracker were People, Government, Trump, COVID, Police, Cases, Etc.; these keywords and entities as shown in Fig. 18.

4.3 Sentiment Analysis

For sentiment analysis on the selected Tracker, we have presented here how BlogTracker can allow analysts to have a user-friendly visualization of various Likert scale emotions variables. We considered some key human emotions for which we compute emotion distribution; these emotion scales are named; Joy, Fear, Anger, Trust, Disgust, Surprise, Anticipation, and Sadness Fig. 19. Gauging the various blog posts in the COVID-19 Tracker lets us see which blog post and date show each Likert scale emotion. We also have a feature on BlogTracker that measures the toxicity and emotion score of various blog posts and the author. We visualize these using the radar chart Fig. 20 and allow analysts to compare these scores on the radar chart. Analysts can also compare two different blog posts to see how they perform. Using the radar chart Fig. 18 visualization, we observed that the emotions of the two compared blog posts are negative emotions symbolic of the effects of a pandemic and reaction towards a pandemic.

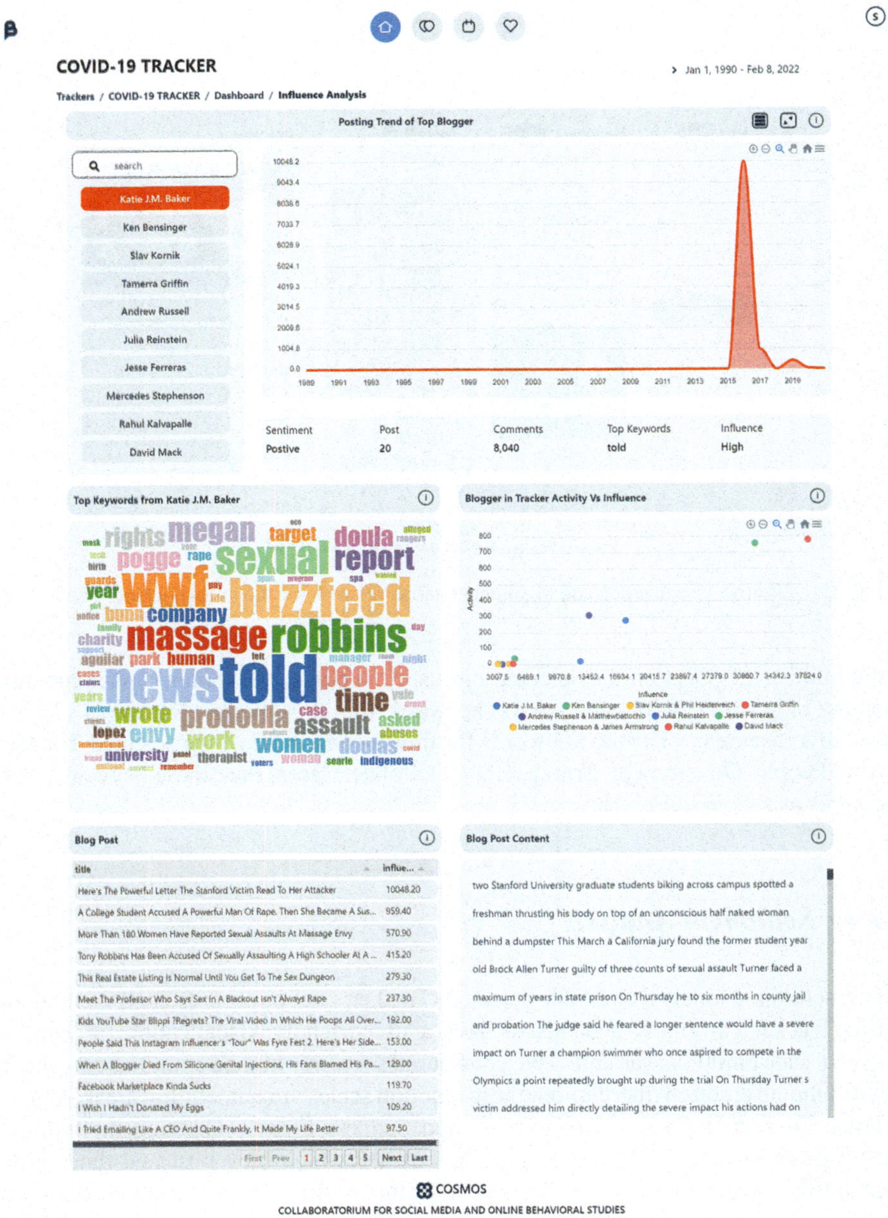

Fig. 18 Influence analysis dashboard, keyword cloud, influence activities, and statistics of a selected blogger

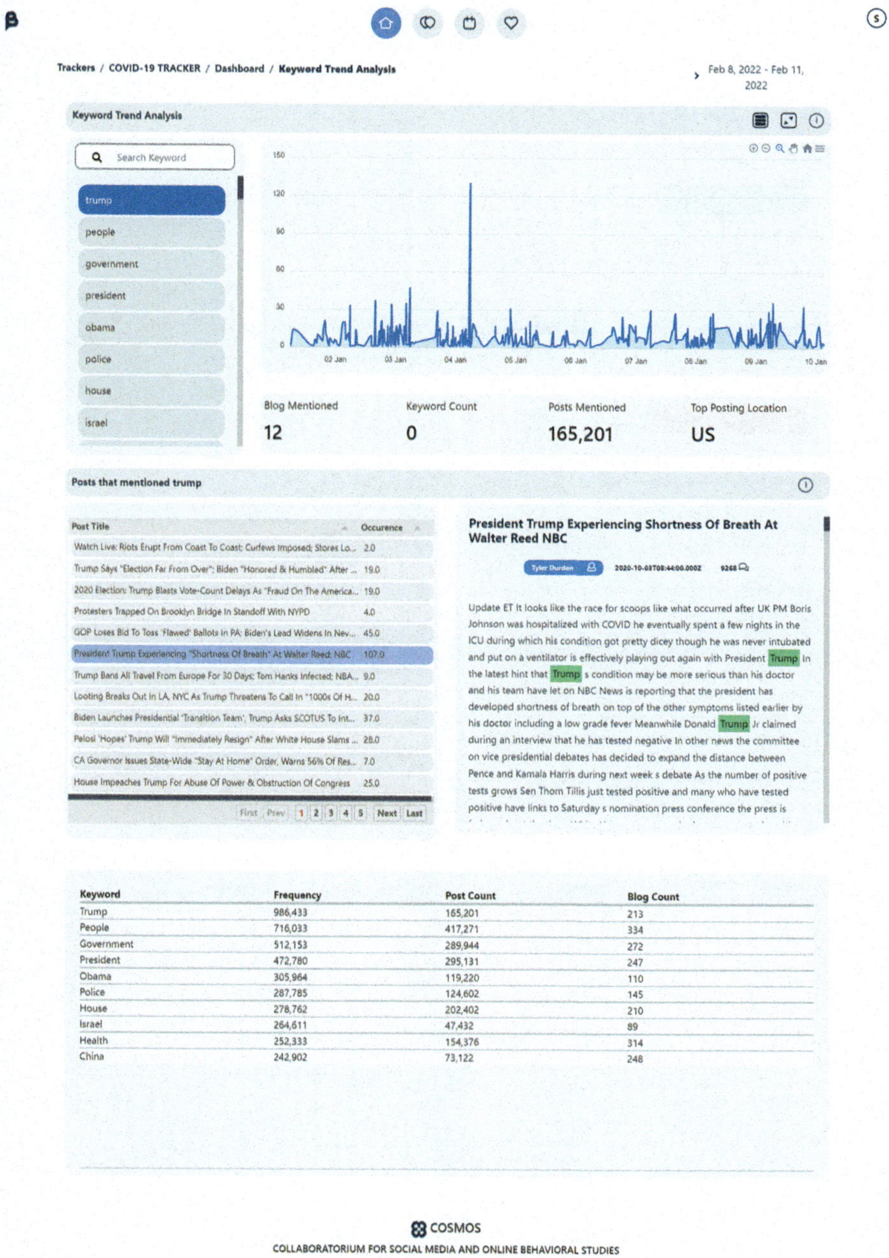

Fig. 19 Keyword trend visualization showing top keywords and posts mentioned

Fig. 20 Visualization showing a comparison of selected keywords

4.4 Narrative Analysis

The narrative capability of BlogTracker enables analysts to see how each entity and the respective blog title and narratives that the blog post author is influencing can be visualized easily. We used our narrative extraction model described in Fig. 3) adapted after Hussain et al. [9]. These entities, narrative and associated blog posts were then visualized using our intuitive developed narrative visualization interface as presented (Fig. 21) and Hussain et al. [9]. We also provide an analyst with the ability to group these entities, merge narratives, and edit these narratives. Also, the keywords of these narratives were highlighted in the blog post, and Analysts can also sort these narratives and entities by criteria like relevance, date, and alphabetical order were highlighted in the blog post, and Analysts can also sort these narratives and entities by criteria like relevance, date, and alphabetical order.

4.5 Influence Analysis

We measured how influence is formed for the selected Tracker; our BlogTracker application was used to compute the top ten bloggers that influence the discussion of the COVID-19 Tracker. We then measured and visualized their influence and the posting pattern of these top bloggers that are dominant influencers within the selected Tracker. We provide a section that shows the statistics like how strong their influence is, the top keyword used by these bloggers, total comments the post by the blogger has generated over the years as seen in Fig. 22 which also shows the statistics of a selected blogger 'Katie J.M Baker', in the COVID-19 Tracker, has a positive sentiment attached to her blog sites, her blogs over 8000 comments we also show the top keywords Fig. 23. Our tool also allows analysts to compare multiple bloggers' influences and visualize these bloggers' statistics.

4.6 Top Keyword Trends

We also used our keyword trend feature capability in this case study, and we were able to follow dominant keywords in the selected COVID-19 Tracker. According to the statistics computed by the BlogTracker tool, the top posting location for the selected keyword 'Trump' in Fig. 19 is seen to be the United States, with over 165,000 posts mentioning the selected keyword. Figure 19 also shows the trend of the selected keyword in the COVID-19 Tracker. The keyword trend analysis feature in the BlogTracker tool is also able to highlight various blog posts in which the selected keyword in Fig. 19 occurs dominantly. It also has a side panel of the blog post content and the keyword highlighted with this blog post content. This helps the

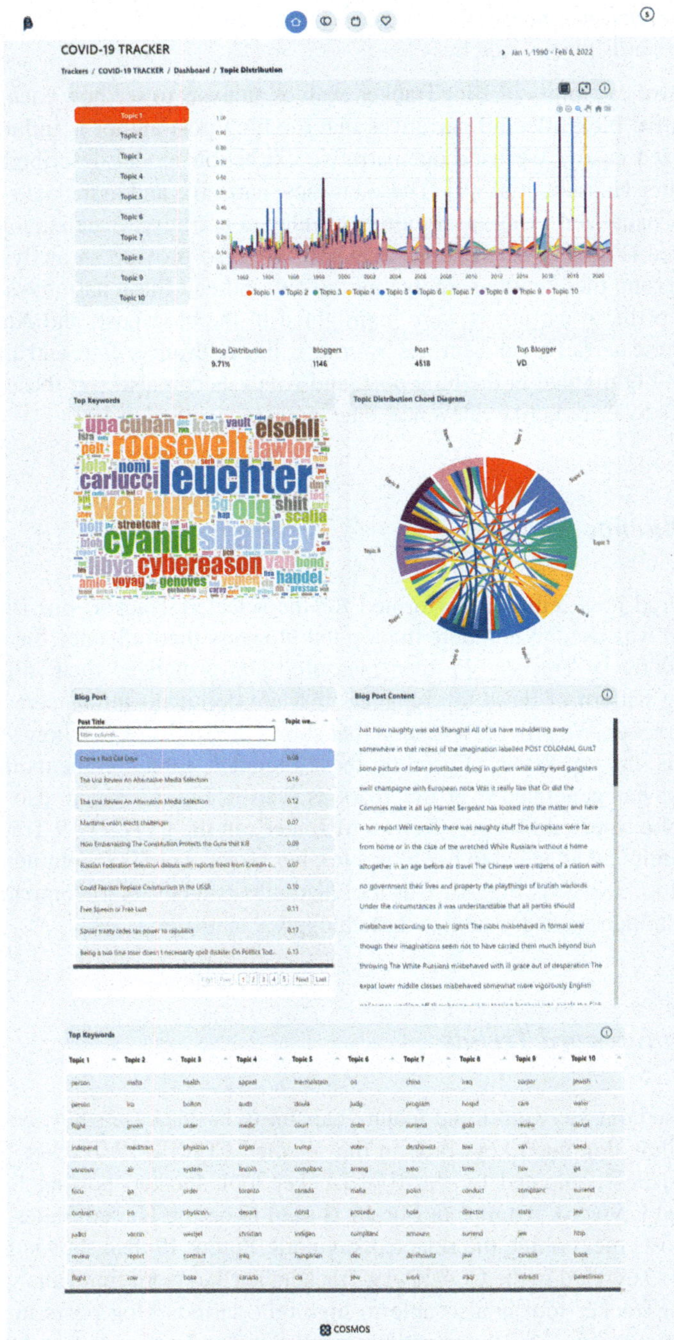

Fig. 21 Visualization and statistics of topic distribution of COVID-19 tracker

Analyst to emphasize these sets of words, with Fig. 20 showing a comparison trend chart of keywords placed side by side.

4.7 *Topic Distribution Analysis*

Furthermore, we showcase the capability of the BlogTracker tool in extracting and analyzing engagement topics. We generated the top ten topics and then tabulated them and their associated keywords to help analysts have in-depth insights into what dominant words were in each topic. Figure 21 shows the visualization of the topics distribution trend chart, while a card on Fig. 21 also shows the keywords word cloud of a selected topic and a chord diagram showing topic overlaps of how each blog post mentioned these keywords. In the selected topic, Topic 1 from the COVID-19 Tracker, we see the statistics show that the distribution of the total blogs in this topic is 9.71% of the total blogs in this Tracker, with a total of 1146 bloggers and an associated 4518 posts, we also aptly see that the top blogger in this Topic 1 is VD. Furthermore, on careful analysis of the 4518 posts in Topic 1. Our BlogTracker tool was also able to group these dominant keywords under their respective topics in a table format with words like Trump, vaccine, Canada, and other trending words extracted from the various blog posts under study.

5 Conclusion and Future Works

In this paper, we focused on the capability of the BlogTracker tool to meet the demand of analysts who have had limited resources in studying the blogosphere using a centralized tool, where other applications capable of analyzing various aspects of social and behavioral science may be unavailable. We also show how BlogTracker can find narratives and make the visualization interactive. We also demonstrated how to use BlogTracker to extract important topics, sentiments, and other opinions and user emotion activities towards a blog post. To show the BlogTracker tool's capabilities in crunching a large dataset and helping analysts fast-track research and behavioral studies, we used the COVID-19 Tracker and examined 852,106 blog posts from over 380 unique blogs sites. We were able to show some salient features of the BlogTracker application https://btracker.host.ualr.edu. Our work also solved the need for ready and easy access tools backed by years of research and continuous learning.

Future works will entail tracking the network of activities around deviant behaviors and how the blogosphere is affected by bots posting hyperlinks to various misinformation blog sites. We hope to incorporate capabilities like this into BlogTracker to further extend the system's capabilities by understanding how narratives form and how bots may have influenced narratives and many other emerging issues.

Acknowledgments This research is funded in part by the U.S. National Science Foundation (OIA-1946391, OIA-1920920, IIS-1636933, ACI-1429160, and IIS-1110868), U.S. Office of Naval Research (N00014-10-1-0091, N00014-14-1-0489, N00014-15-P-1187, N00014-16-1-2016, N00014-16-1-2412, N00014-17-1-2675, N00014-17-1-2605, N68335-19-C-0359, N00014-19-1-2336, N68335-20-C-0540, N00014-21-1-2121, N00014-21-1-2765, N00014-22-1-2318), U.S. Air Force Research (FA9550-22-1-0332), U.S. Army Research Office (W911NF-20-1-0262, W911NF-16-1-0189), U.S. Defense Advanced Research Projects Agency (W31P4Q-17-C-0059), Arkansas Research Alliance, the Jerry L. Maulden/Entergy Endowment at the University of Arkansas at Little Rock, and the Australian Department of Defense Strategic Policy Grants Program (SPGP) (award number: 2020-106-094). Any opinions, findings, and conclusions or recommendations expressed in this material are those of the authors and do not necessarily reflect the views of the funding organizations. The researchers gratefully acknowledge the support.

References

1. Agarwal, N., Liu, H., Tang, L., Philip, S.Yu.: Identifying the influentials bloggers in a community. In: Proceedings of the International Conference on Web Search and Data Mining (WSDM '08). Association for Computing Machinery, New York, NY, USA, 207–218. (2008). https://doi.org/10.1145/1341531.1341559
2. Akinnubi, A., Agarwal, N., Stine, Z., Oyedotun, S.: Analyzing online opinions and influence campaigns on blogs using Blogtracker. In: Proceedings Of The 2021 IEEE/ACM International Conference On Advances In Social Networks Analysis And Mining. pp. 309–312 (2021)
3. Al-Ani, B., Mark, G., Chung, J., Jones, J.: The Egyptian blogosphere: A counter-narrative of the revolution. In: Proceedings of the ACM 2012 conference on Computer Supported Cooperative Work (2012)
4. Bandeli, K., Hussain, M., Agarwal, N.: A Framework Towards Computational Narrative Analysis On Blogs. (2020)
5. Bansal, S.: Beginners guide to topic modeling in python and feature selection (2022). https://www.analyticsvidhya.com/blog/2016/08/beginners-guide-to-topic-modeling-in-python/
6. Barron, A., Huang, J., Spang, R., Dedeo, S.: Individuals, institutions, and innovation in the debates of the french revolution. PNAS. **115**, 4607–4612 (2018)
7. Hossein, H., Beneki, C., Unger, S., Mazinani, M.T., Yeganegi, M.R.: Text Mining in Big Data Analytics Big Data and Cognitive Computing 4, no. 1: 1. (2020). https://doi.org/10.3390/bdcc4010001
8. Hussain, M.N., Bandeli, K., Al-khateeb, S., Agarwal, N.: Analyzing shift in narratives regarding migrants in Europe via the blogosphere. In: CEUR Workshop Proceedings (Vol. 2077). CEUR-WS. (2018)
9. Hussain, M.N., Rubaye, A.I., Bandeli, K.K., Agarwal, N.: Stories from blogs: computational extraction and visualization of narratives. In: (Text2Story)@ECIR, pp. 33–40 (2021)
10. Kolari, P., Finin, T.: Memeta: a framework for multi-relational analytics on the blogosphere. In: AAAI 2006 Student Abstract Program (2006)
11. Peng, S., Wang, G., Xie, D.: Social influence analysis in social networking big data: opportunities and challenges. In: IEEE Network, vol. 31, no. 1, pp. 11–17, 10.1109/MNET.2016.1500104NM. (2017)
12. Pham, M.C., Derntl, Y., Cao, Y., Klamma, R.: Learning analytics for learning blogospheres. Learn. Anal. Learn. Blog. **7558**, 258–267 (2012)
13. Shaik, M., Hussain, M.N., Stine, Z., Agarwal, N.: Developing situational awareness from the blogosphere: an Australian case study. In: The Eleventh International Conference On Social Media Technologies, Communication, And Informatics (2021)
14. Stine, Z., Agarwal, N.: Characterizing the language-production dynamics of social media users. Soc. Netw. Anal. Min. **9**, 60–60 (2019)

15. Singh, V. K., Mahata, D., Adhikari, R.: Mining the Blogosphere from a socio-political perspective. International Conference on Computer Information Systems and Industrial Management Applications (CISIM), Krakow, Poland, pp. 365–370, (2010) 10.1109/CISIM.2010.5643634.
16. Tsirakis, N., Poulopoulos, V., Tsantilas, P., Varlamis, I.: Large scale opinion mining for social, news and blog data. J. Syst. Softw. **127**, 237–248 (2017)

Parallelized Cyber Reconnaissance Automation: A Real-Time and Scheduled Security Scanner

Malek Malkawi and Reda Alhajj

Abstract The extraordinary advancement of technology has increased the importance of achieving the required level of information security, which is still difficult to achieve. Recently, network and web application attacks have become more common, causing confidential data to be stolen by exploiting system vulnerabilities. The CIA Triad Model is broken as a result of this. In this work, with the aim of relieving real-world concerns, we present an enhanced schema for the first feature of the security engine we proposed in the previous paper. It is an automated security scanner based on parallelization for the active information-gathering phase. It supports real-time and scheduled system scans in parallel in the phase of active information gathering based on RESTful API allowing easy integration for real-life cases. With the integration of the message-broker software (RabbitMQ) that originally implemented the advanced message queuing protocol (AMQP), the user has the ability to create instant customized scans and check the related results. These features depend on Celery workers using asynchronous task queue which is reliant on distributed message passing to perform multiprocessing and concurrent execution of tasks. The system can be used by penetration testers, IT departments, and system administrators to monitor their system and grant high security and instant alarms in critical threats. An automated IP and port scanning, service-version enumeration, and security vulnerabilities detection system are the core of the proposed scheme project. The accuracy and efficiency of this technique have been demonstrated through a variety of test cases based on real-world events. The average time of scanning a server and detecting the vulnerabilities has been enhanced by 22.73% to become 1.7 minutes instead of 2.2 minutes. Similarly, the improvement ratio for run time, elapsed time and vulnerability detection are 20.40, 90.80, and 7.70% respectively.

M. Malkawi (✉) · R. Alhajj
Vocational School, Istinye University, Istanbul, Turkey

Department of Computer Engineering, Istanbul Medipol University, Istanbul, Turkey

Department of Computer Science, University of Calgary, Calgary, AB, Canada

Department of Heath Informatics, University of Southern Denmark, Odense, Denmark

S. T. Özyer, B. Kaya (eds.), *Cyber Security and Social Media Applications*, Lecture Notes in Social Networks, https://doi.org/10.1007/978-3-031-33065-0_2

Keywords Security scanner · Cyber reconnaissance · Vulnerabilities · Penetration testing · Automation · Parallelization

1 Introduction

Internet technology has proliferated and progressed at a breakneck pace. The internet connects people's lives, but it also brings with it some threats, such as cyber-attacks [1–3]. That may result in data leaking or business interruption, leading to significant financial losses [4] and breaking the CIA (Confidentiality, Integrity, and Availability) Triad, which is the main principle of information security. Penetration Testing (PenTest) is one of the methods used to discover possible attacks and hidden threats. One of the most important aspects of testing is identifying vulnerabilities in a timely manner [5–8]. Existing vulnerabilities mean that data processed by web apps is not secure and exposed to the existing exploitation scenario [9–11].

Manual testing approaches make it difficult to detect different types of application vulnerabilities. The resources allocated for testing are limited and not only the complexity of modern web applications is increasing but also the number of problems to be solved in terms of the integration among the used technologies. That's why regular and automated vulnerability assessments help to reduce the risks of eventual intrusion and violations of the systems' data integrity, availability, and confidentiality [12–19].

Reconnaissance is a fundamental part of the penetration testing methodology in which a professional penetration tester gathers as much information about a target machine or network as possible prior to conducting a guided test in search of vulnerabilities. This process consists of three phases: fingerprinting, scanning, and enumeration.

In this paper, we are going to propose an parallelized security scanner which is the extended version of our work done in [20]. We provide an API-based automated network scanner including host discovery, open port scanner, service-version enumerator, and vulnerability detection approach using the Network Mapper (Nmap) Scripting Engine (NSE) integrated with VulScan and Nmap-vulners modules databases to find Common Vulnerabilities and Exposures (CVE). VulScan and Vulners are NSE scripts that make use of publicly available services to provide relevant information about security vulnerabilities [21]. Parallel real-time and scheduled system scans are supported in this version. With the integration of RabbitMQ message-broker, which originally developed the advanced message queuing protocol (AMQP), the user can create instant customized scans and retrieve the relevant results. These functionalities rely on Celery workers that perform multiprocessing and concurrent task execution in the main endpoint of our RESTful-API server utilizing an asynchronous task queue.

The project adheres to Python Enhancement Proposal to enrich readability, as well as modules and packaging, in order to achieve the project's purposes in the most efficient manner feasible. In the new version of our project, a distinct schema has been established for each endpoint to optimize the connectivity. Before being processed, Marshmallow which is a Python package that transforms complicated data types into native Python data types and vice versa [22] will check the inbound request for compliance. In this work, we have proposed a solution that can easily be integrated with the current systems to solve the difficulties faced while switching from one tool to another to complete the process of scanning as the first feature of an automated security engine. The primary goal is to close the gap by automating the network scanning and vulnerability assessment processes for small and large networks. At the same time, it gives the opportunity for administrators and researchers to prove that a host is vulnerable to certain attacks or not by automated periodic scans. Moreover, the test case outcomes exceeded our expectations. We nearly covered all essential network architectures based on real-life scenarios. The outcomes were significantly better than any manual method. We were able to quickly identify weaknesses with high accuracy. The improved schema enhanced the run time, elapsed time, and vulnerability detection by 20.40, 90.80, and 7.70%, respectively.

In the next Sect. 2, we demonstrate the literature review and the related work. In Sect. 3, we introduce the methodology and the main components used in the research. Then Sect. 4, we illustrate the details of the implementation and the applied approach. Section 5 shows the results we have obtained and the test cases' scenarios. The following Sect. 6 is the discussion of the work. Finally, the conclusion and future work are explained in Sect. 7.

2 Related Work

Reconnaissance is the process by which a hacker acquires information about a target before launching an attack or penetration test. It is carried out in stages prior to exploiting or patching system vulnerabilities. There are various tools for doing specific parts of this process. However, research towards automatic and comprehensive tools is still in its infancy.

Shah et al. [23] demonstrate traffic accountability and the time required to perform a specific task during active scanning with the Nmap tool, as well as propose solutions for dealing with huge volumes of hosts while preserving network traffic and task time. Nmap is also used by certain research groups only as a penetration tool to test intrusion detection systems' (IDS) performance and discover the related effects [24]. In [25] the authors concentrate on Nmap's operating system recognition feature and use it to acquire information about the target OS by matching IP/TCP stacks with the tool's built-in fingerprint database. In [26] ZMap has been developed which is an internet-scale network scanner that scans single packets quickly. It can be used in conjunction with ZGrab, a sister project that supports stateful application-layer handshakes, grabs banners, and detects Heartbleed vulnerabilities.

Schagen et al. [27] developed an automated network vulnerability scanning method that employs safe patch fingerprinting. This approach automates input detection and differentiates vulnerabilities across non-patched and patched systems quickly. Furthermore, this method allows specialists to scan and safeguard networks efficiently. In order to ensure the safe and ethical execution of patches, inputs that are suspected to be somewhat malicious are denied [27]. Roy et al. [28] developed a Java-based command line application that locates and stores an organization's footprint and looks for sensitive information, such as data repositories, to enhance an organization's vulnerability assessment process [28]. To determine the correlation between attacks and port scans, Panjwani, S. et al. [29] employ an experiential technique. To test the theory, they ran trials for 48 days on two target computers on a widely used subnet.

In [30], Jensen J. Zhaoet al. performed research to evaluate the security of US e-governance sites. Though there were a few best practices in place, such as the usage of SSL and a firewall, areas such as the clarity of confidentiality statements and open port 80 were discovered, and viable steps for strengthening e-governance security were suggested. P. LaRoche et al. [31] presented their research into using generic programming to evolve valid TCP packets, conduct a variety of port scans through an IDS, and remain covert in their activity. In [32], McPherson, J.et al. describe a system called "portvis" that analyzes coarsely detailed security-related network data of a basic sort and visualizes security-related events of interest. The authors in [21] discuss some of the challenges that arise while analyzing vulnerabilities and identifying current web application vulnerabilities, as well as some potential solutions.

The study in [33] seeks to conduct a survey of typical consumer computer devices, such as start devices, in order to identify open ports and, as a result, vulnerabilities that could lead to potential attack targets. In [34], the authors describe a scanner that detects SQL Injection, Cross-Site Scripting (XSS), CRLF Injection, and Open Redirect vulnerabilities in web applications. It also includes a rudimentary port scanner and a web crawler module that aids in the detection of other services operating on the web server.

3 Methodology and Main Components

The fundamental components used to implement the project, as well as the methodological approach in terms of the overall structure, will be detailed in this section.

3.1 Python Programming Language

Python is a high-level programming language with a design philosophy that prioritizes readability. Its features and object-oriented approach are designed to help programmers write clean code for both small and large-scale projects. Besides, it's a beneficial language for cybersecurity since it is capable of doing a wide variety of cybersecurity tasks, such as malware analysis, scanning, and penetration testing [35, 36]. Those are some of the reasons why we have decided to write our project with it.

3.2 Network Scanner (Nmap)

Nmap (Network Mapper) is an open-source network scanning and discovery tool widely used for scanning the state of a target host [37–39], and it provides comprehensive scan types and firewall evasion methods. Each technique can be customized and made to be as noticeable or as inconspicuous as possible. It can be used to find open ports, communication protocols (TCP/UDP), services and versions used on each port, as well as vulnerabilities on a remote device. The data acquired can be used to improve the system in terms of preventing future attacks. Our schema relies on Nmap algorithms to gather the required information regarding the target [38, 40].

3.3 Nmap Scripting Engine (NSE)

One of Nmap's most vital and flexible features is the Nmap Scripting Engine (NSE). It allows users to build simple scripts to automate a wide range of networking functions using a programming language called Lua. Those scripts are run in parallel, with the speed and efficiency that Nmap is customized for. Users can use the built-in scripts of Nmap's library or develop their own to fulfill their specific needs. NSE's key tasks include network discovery, more sophisticated version detection, vulnerability detection, and backdoor detection. It can even be used for vulnerability exploitation [38, 41].

3.4 Flask Micro Web Framework

Flask is a Python-based micro-framework as it does not necessitate the usage of any specific tools or libraries. It doesn't have a database abstraction layer, form validation, or any other components that rely on third-party libraries to do typical tasks. When it comes to designing web applications, Flask gives developers a lot of options. It includes tools, libraries, and mechanisms that allow you to build a web application, but it doesn't impose any dependencies or tell the user how the project should look [42].

3.5 RESTful API

REpresentational **S**tate **T**ransfer is a software architectural style that was developed to help in the design and development of the WWW's architecture. REST establishes a set of guidelines for how an Internet-scale distributed hypermedia system, such as the Web, should be designed. As shown in Fig. 1, REST APIs offer a lot of flexibility, which is one of its biggest benefits. REST can accommodate many sorts of calls, return diverse data formats, and even change fundamentally with the correct implementation of hypermedia because data is not linked to resources or functions. Data and functionality are considered resources in the REST architectural style, and they are accessible via Uniform Resource Identifiers (URIs). The most often used protocol is HTTP. Using such an architecture, making projects more appropriate for scaling up and integrate it with other systems. All of these characteristics contribute to the simplicity, lightness, and speed of RESTful applications [43].

3.6 Celery Task Queue

Celery is a Python-based open-source asynchronous task queue that uses distributed message passing to divide work across multiple threads or processors. Any language can be used to implement the protocol. The input to a task queue is a task, which is a unit of work. Using multiprocessing, eventlet, or gevent, these tasks are carried out concurrently on one or more worker nodes. Celery communicates with clients and workers via messages, with a broker acting as a middleman. RabbitMQ and Redis are the suggested message brokers. To begin a job, the client adds a message to the

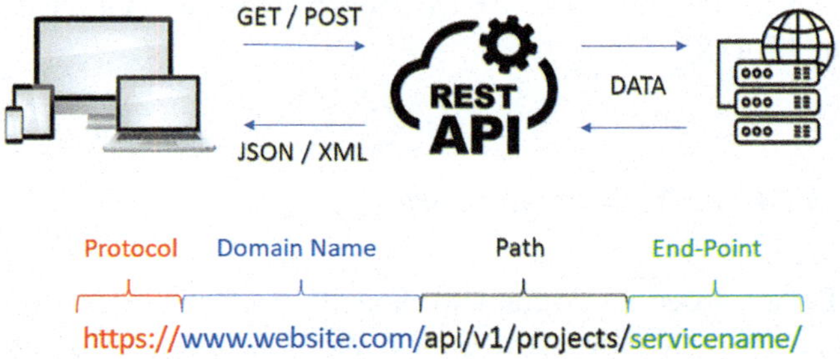

Fig. 1 API architecture

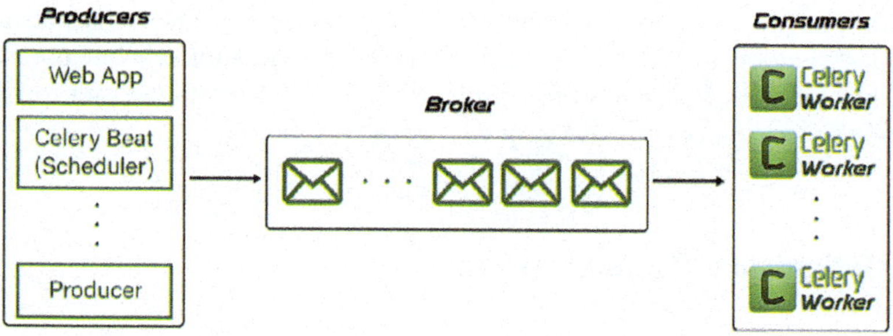

Fig. 2 Celery architecture

queue, which is then delivered to a worker by the broker as shown in Fig. 2. Multiple workers and brokers can be used in a Celery system, allowing for high availability and horizontal scaling. Celery enables scheduling tasks via Celerybeat and real-time operations via Celery workers [44].

3.7 RabbitMQ Message Broker

RabbitMQ is an open-source message broker developed in Erlang programming language, which was designed to implement the Advanced Message Queuing Protocol (AMQP). Message brokers are responsible for transporting messages and providing producers with the opportunity to begin working on a different task as shown in Fig. 2. The sender does not have to wait for the message to be received by the recipient. It is a versatile system that can be used to reduce web application server load and delivery times. Persistence, delivery acknowledgments, publisher confirms, and availability are the capabilities available in RabbitMQ to balance performance and reliability. On a local network, several RabbitMQ servers can be clustered together to form a single logical broker, and queues can be mirrored in the cluster to guarantee that messages are safe even if the hardware fails. RabbitMQ supports messaging via a variety of messaging protocols and has an easy-to-use management user interface that allows the user to monitor the message broker [45–47].

3.8 Flower Monitoring Tool

Flower is a web-based utility that allows the user to keep track of Celery clusters, task progress, details, and worker status. It has a variety of functions to assist the user in comprehending and being aware of what is going on. Some of the features of Flower include real-time monitoring with task progress and history graphs and statistics, as well as a remote control with the ability to view worker status and statistics, shutdown and restart worker instances, and monitor scheduled, reserved, and canceled jobs. It also has broker monitoring, which allows the user to view all not only Celery queues but also graphs of queue length [48].

4 Methods and Implementation

The project has been implemented in the Python 3.9 programming language and the Flask Micro Web-Framework as a RESTful-API Server based on Network Mapper (Nmap) algorithms with Celery and RabbitMQ. This approach gives the chance to integrate the project with any system. We have implemented a main endpoint called "AutoScan" requiring IP, port range, scan type and option input to go over the whole process in an automated and parallelized approach, starting with the host discovery, port scanning, service-version enumeration, and ending with vulnerability detection. The detailed result of the scan will be presented as JSON output. Each function of the active reconnaissance process has been implemented as a separate endpoint as shown in Fig. 3 to increase the accuracy and optimize the schema. Additionally, it allows advanced users to enter their own customized parameters that might be needed. Each function has been explained in the following subsections.

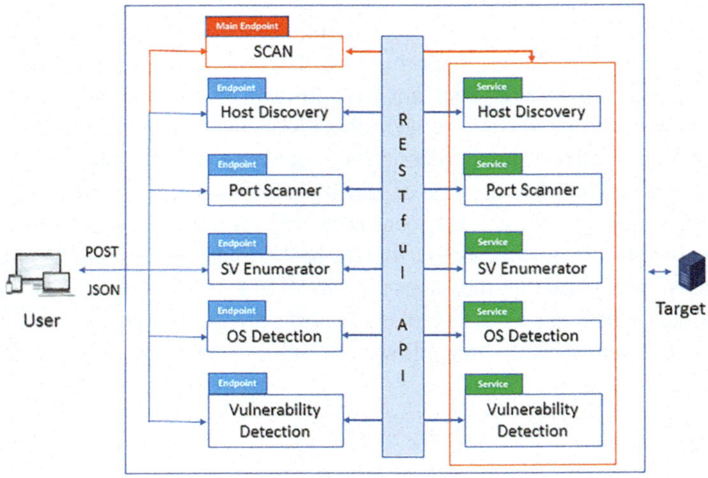

Fig. 3 General structure

4.1 Host Discovery

Host discovery is the first endpoint in our system. "<server-ip>/api/v1/host-check". At this point, the given IP's (target system) status will be checked if it is active or not and the result will be as "Up" or "Down". Accordingly, the system will process to the next step (port scan endpoint) automatically. If the result is down then it will go directly to the final step without wasting time with the other scanning methods.

There are three primary host discovery detection methods as demonstrated in Fig. 4. The first one is ARP detection, which in turn broadcasts ARP query packets throughout the LAN. Using a list scan rather than a PING scan by setting the target host's state to "HOST UP" is the second one. The other detection methods include sending four different types of data packets to determine whether the target host is online; An ICMP timestamp request, an ICMP echo request, a TCP SYN packet to port 443, a TCP ACK packet to port 80, and a TCP SYN packet to port 443.

4.2 Port Scanning

Port scanning "<server-ip>/api/v1/port-scan" is the schema's second endpoint. By default, the system will start scanning the most known 1000 TCP port, unless the user changes the POST request to scan a specific port range with the demanded network protocol, such as TCP, UDP, or both together. Open ports and the related protocol results will be stored temporarily on the server to be prepared and sent to the next endpoint, which is the service and version enumeration step. In the case of having no open ports detected, the system will terminate the process and show the results.

The goal of a port scan is to determine the target port's operational condition by returning the packet's characteristics. Different probe packets are created for

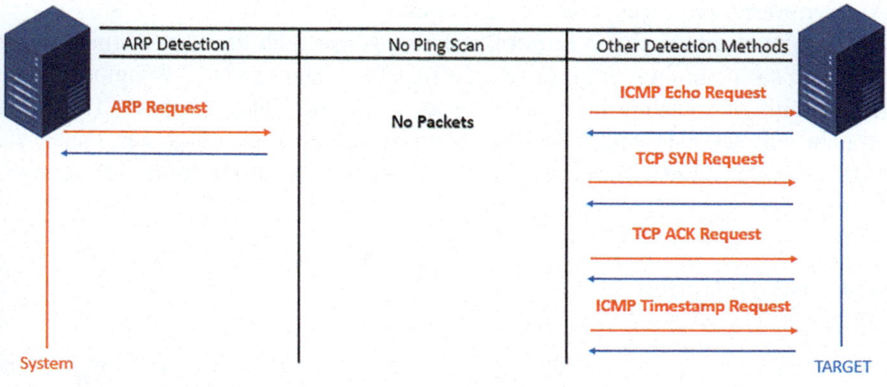

Fig. 4 Host discovery methods

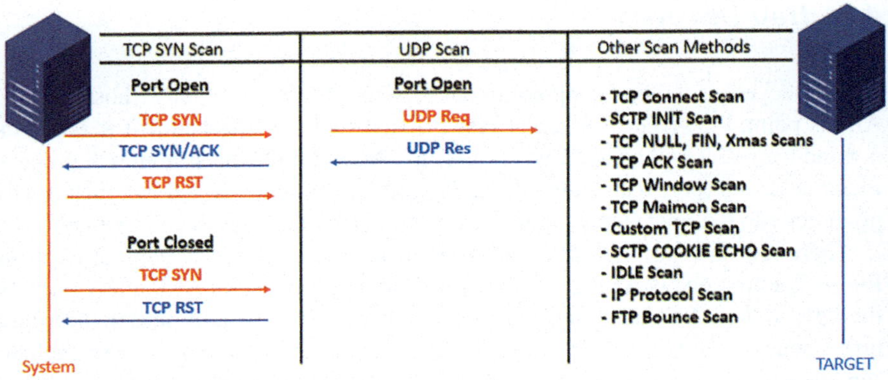

Fig. 5 Port scan methods

different scan types. Nmap first runs a PING probe operation and transmits the necessary probe packets to the target machine's specified port, as shown in Fig. 5. Then it either waits for the probe packet to be re-transmitted or sends a fresh probe packet. Finally, it is dependent on several detection mechanisms and waits for various types of response packets.

4.3 Service and Version Enumeration

At this endpoint "<server-ip>/api/v1/sv-scan", the system enumerates only the banners of the services on the open ports, then the version of each one. This part is the most vital one, as in the next step we are going to detect the related vulnerabilities. Open ports and the related protocol results will be stored temporarily on the server to be prepared and sent to the next endpoint, which is the service and version enumeration step. In the case of having no open port detected, the system will terminate the process and show the results.

The premise behind this form of scanning is to match the scan findings to the database's service fingerprint or service signature. Thousands of common service fingerprints or signature traits are integrated into the Nmap database. The target system will generate and return the necessary information, which can match the port's service and service version, in response to a request for a specific target port.

4.4 OS Detection

To increase the accuracy of the results, we have combined the operating system detection with the previous SV Enumeration endpoint. To identify the operating system, the system uses TCP/IP protocol stack fingerprints. Because there are no mandatory specifications for TCP/IP implementation in some areas in the RFC specification, distinct TCP/IP schemes may have their own unique processing mechanisms. The differences in these characteristics are what Nmap uses to determine the type of operating system. Initially, Nmap runs a sequence generation test, sending six TCP probing packets, and extracting the data fingerprint SEQ/OPS/WIN/T1. It then selects a closed UDP port, a closed TCP port, and an open TCP port for complete TCP/UDP/ICMP detection and fingerprint data extraction, respectively. Finally, it compares the fingerprint attributes of known systems included with Nmap to the findings of the detection as shown in Fig. 6.

4.5 *Vulnerability Detection*

Detecting common vulnerabilities and exposures (CVE) and the possible threats are done at this endpoint via the Nmap Scripting Engine (NSE), and at the end, it returns the final results of the scan. It is divided into three main functions;

- Vulscan: Vulscan is a module that extends Nmap's ability to scan for network vulnerabilities. This is an NSE script that assists in the discovery of vulnerabilities on a single target or a network. This script focuses on service and version detection in order to determine the severity of vulnerabilities on the target computer or network. The module uses the most famous databases of vulnerability which includes all the officially announced security issues. The pre-installed databases are
  - scipvuldb.csv—https://vuldb.com
  - cve.csv—https://cve.mitre.org

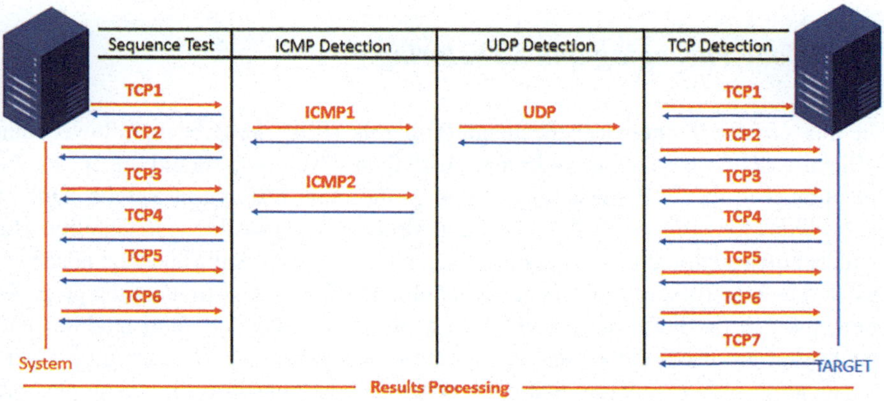

Fig. 6 OS detection

  - securityfocus.csv—www.securityfocus.com/bid
  - xforce.csv—https://exchange.xforce.ibmcloud.com/
  - expliotdb.csv—https://www.exploit-db.com
  - openvas.csv—http://www.openvas.org
  - securitytracker.csv—https://www.securitytracker.com
  - osvdb.csv—http://www.osvdb.org
- Vulners: Vulners is another NSE script that makes use of publicly available services to offer pertinent information about vulnerabilities with more than more than 250GB database size. It is integrated with the nmap libraries in order to improve Nmap's ability to scan for network vulnerabilities.
- Comprehensive: Comprehensive mode makes use of both modules together to detect the highest number possible of vulnerabilities. It checks the results of their return in accordance with their priority in order without any repetitions.

4.6 Main Endpoint: Scan

The main endpoint option, which is one of the major goals of our project, provides both complete automation of the active information reconnaissance process and allows parallel processing of requests received with the support of Celery. What users have to do is just sending the IP with scan option request, and they will be able to review the results when they are ready, without having to wait. The main endpoint will be going through all the phases as shown in Fig. 3. It begins with host discovery, followed by port scanning, detection of operating systems, and lastly vulnerability detection. Depending on the optimization rules we have made, In case of a receiving negative result at any stage, the process will be finalized without initiating the remaining stages by stating the reason.

4.7 Code Packaging and Structuring

In this project, Python Enhancement Proposals (PEP) have been followed that officially define the coding style that all Python developers should adhere to. As mentioned in the PEP, readability is one of the main features of Python. One of the reasons for trying to make the code as structured as possible is readability. The concept of structure involves creating clean code with clear logic and dependencies, as well as how files and folders are ordered in the file-system to meet the project's objectives in the best way possible. To achieve this intent, Python modules and packaging are used. Python modules are one of the most used abstraction layers for dividing code into pieces that hold related data and functions. For example, one layer would be responsible for interacting with user actions, while another would be in charge of low-level data manipulation, and this is done with the "import" and "from" statements. Python has a very clear packaging system, which is essentially a directory extension of the module mechanism. A Python package is any directory that contains an "__init__.py" file that contains all package-wide declarations. Some of the major factors that play a big role in planning and deciding what the project will look like include implementing the functions in the most appropriate modules, data flow through the project, and the relationship between features and functions. Organizing the data and code in modules, encapsulating everything within one directory as a package, using relative paths instead of absolute paths, and choosing file names carefully are the most vital steps taken towards ease of productivity, reliability, and scalability.

4.8 Schema and Data Validation

A unique schema has been developed for each endpoint in the new version of our project to enhance endpoint connectivity. Prior to processing, the incoming request will be verified for compliance and handled appropriately. This is performed using Marshmallow. We have four schemas as follow;

- Host_Check schema requires only IP as a string
- Port_Scan schema requires IP, port, type as a string
- SV_Scan schema requires IP, port, option as a string
- AutoScan schema requires IP, port, type, option as a string

4.9 Scanning Options and Structure

There are now a total of six different vulnerability detection methods available as a result of the improvements we made to the approach. Each one has a flag to enable the script scan using the default set of scripts of the Nmap Scripting Engine.

- Option one with flag one or zero (#10, #11) enumerates the services running on the opened ports, their versions, and then detects vulnerabilities using the "VulScan" script.
- Option two with flag one or zero (#20, #21) follows the same steps as option one, depending on the "Vulners" script.
- Option three with flag one or zero (#30, #31) checks the system using both scripts at the same time.

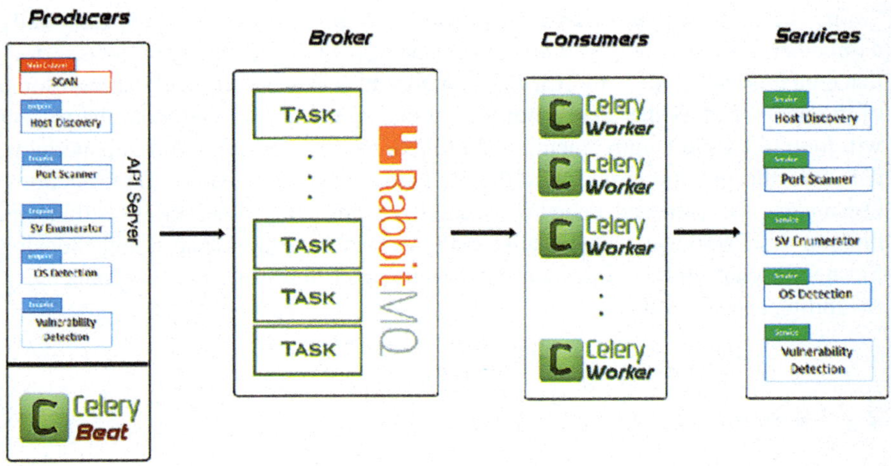

Fig. 7 Project structure

Figure 7 indicates the project structure and the flow of the request. It can be received via the producers, in our case Celery Beat and the RESTful-API. Next, it will be sent to the RabbitMQ broker to be then sent to celery workers according to their availability and processed with the related service.

5 Results and Test Cases

The scheme's performance was assessed using a variety of scenarios based on real-world situations. We tested the project in a total of 20 different environments, covering the majority of the potential topologies. The following are the test cases:

- Metasploitable-2: is a deliberately vulnerable Linux virtual server meant for target practice, training, and exploit testing. Metasploitable2, unlike other vulnerable virtual machines, focuses on operating system and network services layer vulnerabilities rather than custom, vulnerable programs. In terms of vulnerabilities that might be uncovered in a production setting, it has a very similar scenario.
- Bee-box-v1.6: bee-box is a modified Linux Virtual Machine that comes with a buggy web application pre-installed (bWAPP). It gives the chance to investigate almost all vulnerabilities in web applications. bWAPP is a purposefully unsafe web application that is free and open source. It assists web security enthusiasts, developers, and students in identifying and preventing web vulnerabilities. It can

also be beneficial to prepare for effective penetration testing and ethical hacking tasks. The fact that bWAPP has over 100 online vulnerabilities is what sets it apart. It covers all significant known web flaws, as well as all OWASP Top 10 project vulnerabilities. bWAPP is a MySQL database-driven PHP application.

- Metasploitable-1: The Metasploit project is a computer security project helping with penetration testing and IDS signature development by providing knowledge about security flaws. This is the initial version of metasploitable, which is based on an Ubuntu 8.04 server that has been customized. Tomcat 5.5 (with weak credentials), distcc, tikiwiki, twiki, and an earlier mysql are among the insecure programs provided. As previously stated, it is a purposely vulnerable virtual system intended for target practice, training, and exploit testing. It focuses on operating system and network services layer vulnerabilities.
- BTRsys1 and BTRSys2.1: BTRsys project contains intermediate-level boot-to-root susceptible images, with the primary goal of gaining shell access by attacking vulnerable services on the machine, which is a distinct scenario from the real-world probable risks.
- g0rmint: This buggy machine is totally based on a real-world scenario that has been encountered while performing testing for a client's website. It is one of the advanced scenarios in obtaining a limited shell.
- Hades-v1.0.1: Hades is yet another boot2root challenge designed mainly for experienced researchers. Reverse engineering, sploit creation, and a thorough understanding of computer architecture will all be required to exploit this machine successfully. The goal of it is to gradually get more access to the box until the root account is reached.
- vulnVoIP: VulnVoIP is built on the AsteriskNOW distribution, which has some flaws in it. The goal is to track down VoIP users, crack their passwords, and acquire access to the support account voicemail message system. To add a little spice to the mix, this specific distribution is also vulnerable to a well-known exploit that makes it simple to obtain access to the system's root shell. It is one of the few real-life VoIP test environments available.
- VPLE: The Vulnerable Pentesting Lab Environment (VPLE) is a Linux virtual system that is designed to be purposefully vulnerable. This virtual machine can be used for education, tool testing, and standard penetration testing labs. VPLE includes Web-dvwa, Mutillidae, Webgoat, Bwapp, Juice-shop, Security-ninjas, and a Wordpress environment.
- Tr0ll2: The Tr0ll machine has been designed to look and act similarly to the Offensive Security Certified Professional (OSCP) system, and it is prepared to troll the penetration tester at some points. Tr0ll2 is a level of difficulty that is higher than the previous level, Tr0ll1. This is a scenario of intermediate difficulty.
- sick0s1.1: This server provides a clear example of how hacking tactics can be used to infiltrate a network in a secure setting. This virtual computer is

comparable to the ones used in the Offensive Security Certified Professional labs (OSCP). The goal is to get into the network/machine and gain administrative/root access to it.

- MorningCatch: Morning Catch is a VMware virtual machine that demonstrates targeted client-side attacks and post-exploitation, akin to Metasploitable. A website for a bogus seafood firm, a self-contained email infrastructure, and vulnerable Linux and Windows client-side desktop environments can all be found on this virtual server. It also uses WINE to run a few weak Windows apps.
- NETinVM UML 2016: NETinVM is a VMware virtual machine image that exposes a full computer network to the user. As a result, NETinVM can be used to learn about operating systems, computer networks, and security for systems and networks. Furthermore, NETinVM is a VMware virtual machine image that provides a set of User-mode Linux (UML) virtual machines that are ready to use. When the UML virtual machines are started, they build a whole computer network. The Linux operating system is used by all of the virtual computers. It is a full system that displays a real-world network.
- hackxor1: is a webapp testing environment where vulnerabilities must be located and exploited to progress through the system. similar to WebGoat, however, with a focus on realism and challenge. XSS, CSRF, SQLi, ReDoS, DOR, command injection, and other exploits are included. Client attack simulation using HtmlUnit, and realistic vulnerabilities modeled by Google, Mozilla, and other platforms are just a few of the features available.
- Brainpan2, CySCA2014InABox, DonkeyDocker-v1.0, GoatseLinux-1.0-VM, UltimateLAMP-0.2 and w1r3s.v1.0.1: These machines are also designed to have different schemes from real-life ones to practice and test the new penetration testing tools being developed. These vulnerable servers are based on real-world scenarios and similar to OSCP labs.

In each of these test cases, we have tested the enhanced parallelized version of the automation of active reconnaissance implemented in this project and compared it with the previous one [20]. The testing includes all scanning steps; host discovery, port scanning, service-version enumeration, and vulnerability detection with six distinct options. In Table 1, we can see the time elapsed in minutes for each step of the process for each test case of the schema proposed in [20]. And the second table (Table 2) shows the comparison between the enhanced version and the previous work in terms of the scanning time with the enhancement ratio. The vulnerabilities found in each test case with the related improvement ratio are shown in Table 3. The last table (Table 4) shows how long it took to run the new scanning options that were added to the schema.

Table 1 Results of previous version test cases for each endpoint

Virtual Machine Name	Time Elapsed in Minutes: Host Discovery	Port Scan	S.V. Enum. Opt#1	S.V. Enum. Opt#2	S.V. Enum. Opt#3	Results: Total Open Open	S.V. Enum. Opt#1[1]	S.V. Enum. Opt#2[1]	S.V. Enum. Opt#3[1]	OS Detection Detection
Metasploitable2	0.025	0.316	5.020	2.718	5.051	30	41692	417	42109	100
Bee-box_v1.6	0.025	0.325	3.835	2.444	3.877	19	3405	544	3949	100
Brainpan2	0.025	0.287	3.183	3.192	3.191	2	6	9	15	100
Metasploitable1	0.025	0.403	4.127	2.720	4.143	13	19794	403	20197	100
BTRsys1	0.023	0.357	0.487	0.358	0.526	3	594	130	724	100
BTRSys2.1	0.115	0.407	0.596	0.281	0.625	3	606	199	805	100
CySCA2014InABox	0.117	0.490	3.98	2.882	4.194	17	52011	395	52406	100
DonkeyDocker_v1.0	0.026	0.336	0.397	0.183	0.391	2	500	150	650	100
g0rmint	0.025	0.302	0.385	0.192	0.414	2	542	199	741	100
Hades_v1.0.1	0.023	0.481	1.872	0.079	0.180	2	370	27	397	100
GoatseLinux_1.0	0.025	0.359	3.977	3.309	3.994	9	5325	228	5553	100
vulnVoIP	0.024	0.398	3.239	2.708	3.244	8	4329	110	4439	100
VPLE	0.029	0.347	1.212	0.291	1.081	8	1249	245	1494	100
Tr0ll2	0.026	0.384	0.559	0.259	0.599	3	373	23	396	100
sick0s1.1	0.039	1.885	0.489	0.292	0.531	3	399	33	432	100
MorningCatch	0.121	0.593	0.867	0.235	0.883	6	1524	34	1558	100
NETinVM 2016	0.081	0.304	0.306	0.179	0.267	2	0	0	0	100
hackxor1	0.038	2.714	0.311	0.183	0.386	1	587	0	587	100
UltimateLAMP-0.2	0.037	0.370	0.276	0.287	0.304	1	35	31	66	100
w1r3s.v1.0.1	0.130	1.065	0.991	0.406	0.890	4	3450	62	3512	100
MIN	**0.023**	**0.287**	**0.276**	**0.079**	**0.180**	**1**	**0**	**0**	**0**	**100**
AVG	**0.049**	**0.606**	**1.806**	**1.160**	**1.738**	**7**	**6840**	**162**	**7002**	**100**
MAX	**0.130**	**2.714**	**5.020**	**3.309**	**5.051**	**30**	**52011**	**544**	**52406**	**100**

[1]The numbers show the amount of vulnerabilities found after service-version enumeration.

Note: The bold part is the min-avg-max analysis of these results

Table 2 Test cases based previous and enhanced version scanning time comparison **(Time Elapsed in Minutes)**

	Auto Scan - Option#1			Auto Scan - Option#2			Auto Scan - Option#3		
Virtual Machine Name	Previous Version	Enhanced Version[1]	ENH Ratio	Previous Version	Enhanced Version[1]	ENH Ratio	Previous Version	Enhanced Version[1]	ENH Ratio
Metasploitable2	5.362	4.233	21.05%	3.060	2.773	9.36%	5.393	4.334	19.63%
Bee-box_v1.6	4.185	3.711	11.33%	2.794	3.167	-13.37%	4.227	3.767	10.87%
Brainpan2	3.495	3.574	-2.25%	3.504	3.548	-1.27%	3.503	3.588	-2.43%
Metasploitable1	4.554	1.534	66.31%	3.147	0.635	79.83%	4.570	1.567	65.71%
BTRsys1	0.868	0.710	18.21%	0.738	0.563	23.67%	0.906	0.757	16.44%
BTRSys2.1	1.117	0.675	39.60%	0.802	0.538	32.96%	1.146	0.703	38.67%
CySCA2014InABox	4.592	4.070	11.37%	3.489	3.173	9.06%	4.801	4.202	12.48%
DonkeyDocker_v1.0	0.759	0.621	18.13%	0.545	0.534	1.97%	0.753	0.652	13.40%
g0rmint	0.712	0.627	11.95%	0.518	0.495	4.52%	0.740	0.640	13.59%
Hades_v1.0.1	2.377	0.495	79.17%	0.583	0.453	22.41%	0.685	0.515	24.81%
GoatseLinux_1.0	4.360	2.979	31.67%	3.692	3.137	15.04%	4.377	3.208	26.71%
vulnVoIP	3.662	3.385	7.54%	3.131	3.223	-2.94%	3.667	3.575	2.50%
VPLE	1.588	1.041	34.42%	0.666	0.406	39.09%	1.457	0.976	33.03%
Tr0ll2	0.969	0.632	34.77%	0.669	0.494	26.20%	1.010	0.623	38.33%
sck0s1.1	2.414	2.200	8.86%	2.216	2.080	6.13%	2.455	2.216	9.73%
MorningCatch	1.581	0.995	37.02%	0.948	0.716	24.51%	1.597	1.043	34.68%
NETinVM 2016	0.691	0.502	27.23%	0.564	0.518	8.09%	0.652	0.523	19.81%
hackxor1	3.063	2.833	7.51%	2.935	2.672	8.97%	3.138	1.137	63.76%
UltimateLAMP-0.2	0.683	0.667	2.36%	0.694	0.578	16.79%	0.711	0.645	9.20%
w1r3s.v1.0.1	2.186	1.920	12.16%	1.601	1.641	-2.54%	2.084	1.872	10.16%
MIN	**0.683**	**0.495**	**27.56%**	**0.518**	**0.406**	**21.70%**	**0.652**	**0.515**	**21.03%**
AVG	**2.461**	**1.870**	**24.00%**	**1.815**	**1.567**	**13.65%**	**2.394**	**1.827**	**23.66%**
MAX	**5.362**	**4.233**	**21.05%**	**3.692**	**3.548**	**3.90%**	**5.393**	**4.334**	**19.63%**
Total Run Time	**49.215**	**37.405**	**24.00%**	**36.297**	**31.344**	**13.65%**	**47.870**	**36.543**	**23.66%**
Total Elapsed Time	**49.215**	**4.233**	**91.40%**	**36.297**	**3.548**	**90.22%**	**47.870**	**4.334**	**90.95%**

[1]The enhanced version is fully automated and parallelized (ENH: Enhancement)
Note: The bold part is the min-avg-max analysis of these results

Table 3 Test cases based previous and enhanced version found vulnerabilities comparison **(Vulnerabilities Found)**

	Auto Scan - Option#1			Auto Scan - Option#2			Auto Scan - Option#3		
Virtual Machine Name	Previous Version	Enhanced Version[1]	ENH Ratio	Previous Version	Enhanced Version[1]	ENH Ratio	Previous Version	Enhanced Version[1]	ENH Ratio
Metasploitable2	41692	41851	0.38%	417	427	2.40%	42109	42278	0.40%
Bee-box_v1.6	3405	3118	-8.43%	544	544	-	3949	3662	-7.27%
Brainpan2	6	6	-	9	10	11.11%	15	16	6.67%
Metasploitable1	19794	19953	0.80%	403	427	5.96%	20197	20380	0.91%
BTRsys1	594	594	-	130	136	4.62%	724	730	0.83%
BTRSys2.1	606	606	-	199	205	3.02%	805	811	0.75%
CySCA2014InABox	52011	59079	13.59%	395	335	-15.19%	52406	59414	13.37%
DonkeyDocker_v1.0	500	500	-	150	159	6.00%	650	659	1.38%
g0rmint	542	542	-	199	205	3.02%	741	747	0.81%
Hades_v1.0.1	370	370	-	27	25	-7.41%	397	395	-0.50%
GoatseLinux_1.0	5325	5325	-	228	228	-	5553	5553	-
vulnVoIP	4329	4329	-	110	110	-	4439	4439	-
VPLE	1249	1249	-	245	245	-	1494	1494	-
Tr0ll2	373	373	-	23	23	-	396	396	-
sick0s1.1	399	465	16.54%	33	79	139.39%	432	544	25.93%
MorningCatch	1524	1613	5.84%	34	128	276.47%	1558	1741	11.75%
NETinVM 2016	0	0	-	0	6	-	0	6	-
hackxor1	587	704	19.93%	0	0	-	578	704	21.80%
UltimateLAMP-0.2	35	86	145.71%	31	109	251.61%	66	195	195.45%
w1r3s.v1.0.1	3450	4128	19.65%	62	205	230.65%	3512	4333	23.38%
MIN	**0**	**0**	-	**0**	**0**	-	**0**	**6**	-
AVG	**6840**	**7245**	**5.92%**	**162**	**180**	**11.33%**	**7001**	**7425**	**6.05%**
MAX	**52011**	**59079**	**13.59%**	**544**	**544**	-	**52406**	**59414**	**13.37%**
Total	**136791**	**144891**	**5.92%**	**3239**	**3606**	**11.33%**	**140021**	**148497**	**6.05%**

Note: The bold part is the min-avg-max analysis of these results

Table 4 New scanning options time results

Virtual Machine Names	Time Elapsed in Minutes — Auto Scan (New Scanning Options) Option#11	Option#21	Option#31
Metasploitable2	4.252	2.818	4.290
Bee-box_v1.6	3.772	3.176	3.782
Brainpan2	3.574	3.531	1.419
Metasploitable1	1.522	0.669	1.581
BTRsys1	0.732	0.581	0.764
BTRSys2.1	0.696	0.521	0.715
CySCA2014InABox	3.906	3.170	4.046
DonkeyDocker_v1.0	0.641	0.520	0.668
g0rmint	0.632	0.553	0.662
Hades_v1.0.1	0.456	0.442	0.541
GoatseLinux_1.0	2.955	2.922	2.986
vulnVoIP	3.416	3.288	3.591
VPLE	1.072	0.392	1.163
Tr0ll2	0.611	0.511	0.637
sick0s1.1	2.176	2.068	2.219
MorningCatch	1.002	0.726	1.040
NETinVM 2016	0.487	0.553	0.500
hackxor1	2.791	2.761	2.829
UltimateLAMP-0.2	0.659	0.596	0.646
w1r3s.v1.0.1	1.846	1.692	1.897
MIN	**0.456**	**0.392**	**0.500**
AVG	**1.860**	**1.574**	**1.799**
MAX	**4.252**	**3.531**	**4.290**
Total Run Time	**37.199**	**31.489**	**35.976**
Total Elapsed Time	**4.252**	**3.531**	**4.290**

Note: The bold part is the min-avg-max analysis of these results

6 Discussion

The evaluation process detailed in the preceding section demonstrates the project's efficiency and outstanding performance in a variety of assessment scenarios. Users can use six different detection methods to check the state of the host (up/down), check the open ports, locate exposed service and version information, and lastly investigate any possible risks, threats, and vulnerabilities in an automated and parallelized manner. The minimum, average, and maximum values of the time elapsed for each endpoint test case are shown in Fig. 8.

In our approach, scanning option#10 relies on the Nmap Scripting Engine's Vulners module, which provides the common vulnerabilities and exposures reference numbers related to the enumerated service and version information, as well as the exploit priority. Option#20 is more detailed and based on the VulScan script, which gives a significantly longer list that includes not only the CVE reference

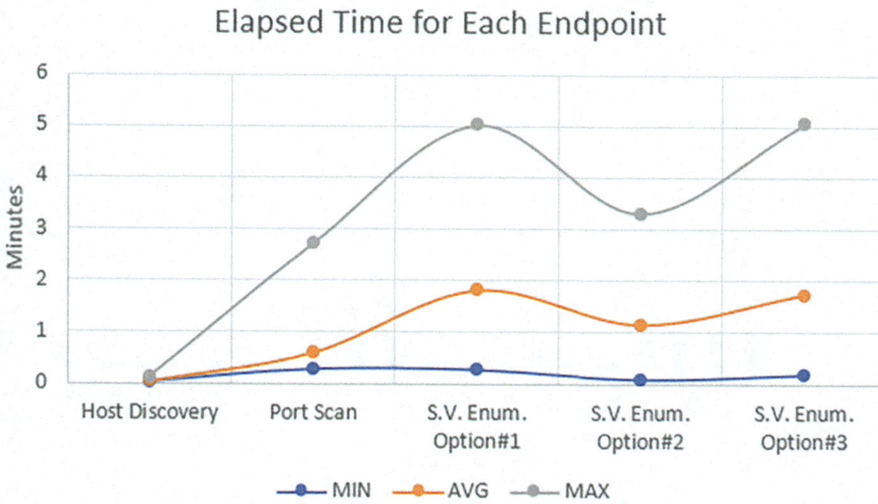

Fig. 8 Total time for each process

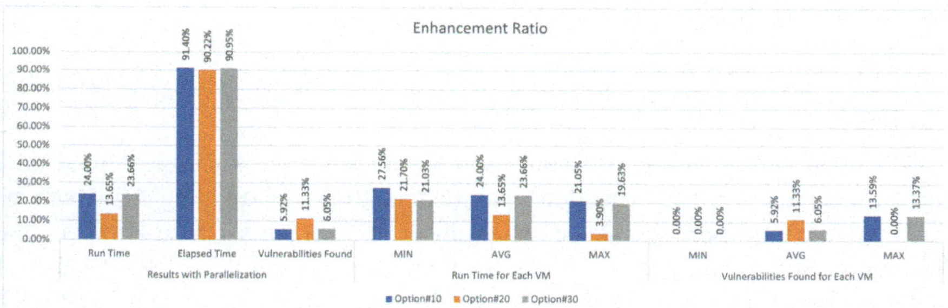

Fig. 9 Enhancement ratio for each option from different aspects

number but also additional potential risks classified by various types of reference numbers. As we mentioned in the methods section, this module is integrated with the world's most well-known eight vulnerability databases, which cover nearly every possible risk related to the organization's technology. The final comprehensive method (option#30) combines both scripts and returns the unique findings. For all options, we can enable the flag by converting the one to zero to lunch the NSE default scripts. Figure 9 demonstrates the enhancement ratio for each strategy in terms of run time, elapsed time, and the number of detected vulnerabilities. The parallelization system used in the scheme accelerated the process significantly. It reduced the total elapsed time to scan the 20 machines from 44.46 to 4.03 minutes, which means a 90.80% improvement. The total run time without parallelization has been enhanced by 20.40% from 44.46 to 35.09 minutes as well. We have detected 98998 vulnerabilities in the current version, while it was 93350 in the previous version. We were able to achieve a 7.70% enhancement. Similarly, the average time

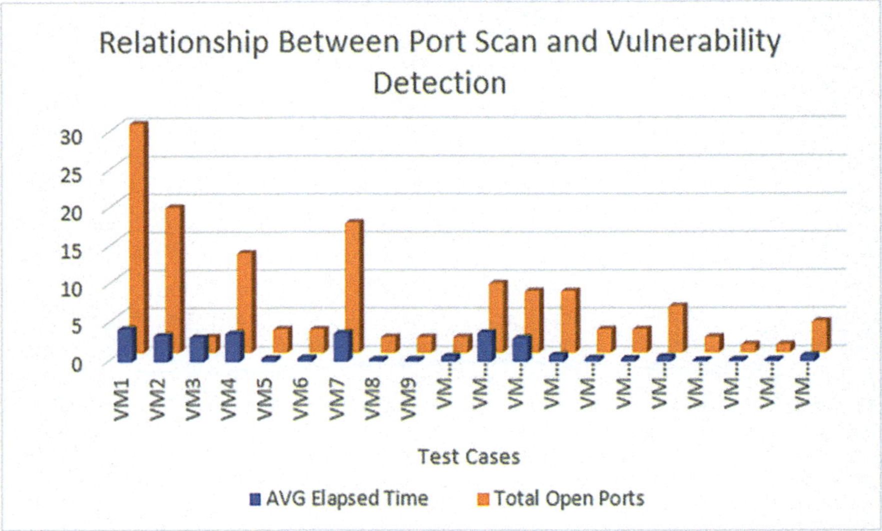

Fig. 10 Port scan and vuln. detection relationship

to scan a server for vulnerabilities was lowered from 2.2 to 1.7 minutes, a 22.73% reduction.

The relationship between the total open port number and the total elapsed time for the enumeration and detection operation is shown in Fig. 10. As it can be expected, the time increases as the total number of open ports grows. However, the increment in time is only about 12 seconds for each port regardless of the number of vulnerabilities. Castiglione et al. [49] indicated that performing a manual approach would take one minute per vulnerability. For example, our system was able to detect 42109 vulnerabilities only in 5 minutes for Metasploitable2 environment. The best case in the increment ratio was 6 seconds and the worst case was 1.5 minute in the test cases. This emphasises the stability and proves the efficiency of the suggested schema in terms of its high processing speed.

Last but not least, the suggested technique will play an important role in system security by allowing companies and system administrators to scan their systems on a frequent and regular basis with the minimal resources they have. Furthermore, this allows for rapid updates to newly discovered software vulnerabilities. Our scanner is used to demonstrate how simple it is to scan a complicated enterprise-grade web application. Attempting to enumerate the service-version and vulnerability information from all ports without making any optimization rule to enhance the performance of the operation is the most common mistake made by other researchers.

7 Conclusion and Future Work

Protecting the CIA (Confidentiality, Integrity, and Availability) Triad, which is at the heart of information security, has become an increasingly difficult task as technology advances. Penetration testing is a method of detecting hidden threats and possible attacks. Manual testing and vulnerability detection procedures, on the other hand, take a long time and make it difficult to uncover various sorts of application faults in a wide network. As a result, regular and automated vulnerability assessments can help decrease the risks of eventual penetration, as well as data integrity, availability, and confidentiality violations, while also promptly discovering flaws.

The goal of this research article is to provide a generic and optimized approach for automating the process using the available and dynamic vulnerability and exposure databases. We've made it apparent that our technique is primarily focused on automating the network scanning and vulnerability detection. In this paper, we have enhanced the proposed schema in the previous paper for the active information-gathering phase. The project is an API-based automated and parallelized security scanner including IP and port scanner, a service-version enumerator, and a vulnerability detection system. In this approach, the Network Mapper is used to collect information with high accuracy using the rules in our schema. Moreover, the work has been developed as a RESTful-API server for easy integration in real-world scenarios, allowing users to scan and protect their networks more rapidly while remaining completely scalable and responsive to their needs and growth.

In the phase of cyber reconnaissance automation, it supports parallel real-time and scheduled system scans. The user can create instant customized scans and view the relevant results with the integration of message-broker software (RabbitMQ) that originally implemented the advanced message queuing protocol (AMQP). These features rely on Celery workers by using asynchronous task queue to perform multiprocessing and concurrent task execution in the main endpoint. The project follows PEP to increase readability, and modules and packaging to meet the project's objectives in the best way possible. To improve endpoint connectivity, a specific schema has been designed for each endpoint in the new version of our project. The inbound request will be checked for compliance using Marshmallow before being processed.

Furthermore, the test case outcomes were satisfactory and beyond our expectations. Based on real-life events, we nearly covered all of the conceivable fundamental conditions. In comparison to a more manual procedure, the results were drastically different. We were able to detect vulnerabilities with high accuracy in a short period of time. We improved the average time of scanning a server and detecting the vulnerabilities by 22.73% to become 1.7 minutes in the new version of the schema, as well as the run time, elapsed time, and vulnerability detection by 20.40, 90.80, and 7.70% respectively.

In the future, our schema will be upgraded to cover the remaining parts of security scanning, as this is just an updated version of the first aspect of the security engine we aim to develop. The next steps will be to build an alarm system, add unit tests for the code, and connect the project to a database. Following this, the project will then be used in a real-world environment to ensure the efficiency and performance of the technique.

References

1. Gamundani, A.M., Nekare, L.M.: A review of new trends in cyber attacks: a zoom into distributed database systems. In: 2018 IST-Africa Week Conference (IST-Africa), p. 1. IEEE, Piscataway (2018)
2. Arnaldy, D., Perdana, A.R.: Implementation and analysis of penetration techniques using the man-in-the-middle attack. In: 2019 2nd International Conference of Computer and Informatics Engineering (IC2IE), pp. 188–192. IEEE, Piscataway (2019)
3. Zhu, N., Chen, X., Zhang, Y.: Construction of overflow attacks based on attack element and attack template. In: 2011 Seventh International Conference on Computational Intelligence and Security, pp. 540–544. IEEE, Piscataway (2011)
4. Kang, S., Qiaozhong, D., WeiQiang, Z.: Space information security and cyberspace defense technology. In: 2013 IEEE International Conference on Green Computing and Communications and IEEE Internet of Things and IEEE Cyber, Physical and Social Computing, pp. 1509–1511. IEEE, Piscataway (2013)
5. Daria, G., Massel, A.: Intelligent system for risk identification of cybersecurity violations in energy facility. In: 2018 3rd Russian-Pacific Conference on Computer Technology and Applications (RPC), pp. 1–5. IEEE, Piscataway (2018)
6. Markov, A., Fadin, A., Tsirlov, V.: Multilevel metamodel for heuristic search of vulnerabilities in the software source code. Int. J. Control Theory Appl. **9**(30), 313–320 (2016)
7. Pechenkin, A.I., Lavrova, D.S.: Modeling the search for vulnerabilities via the fuzzing method using an automation representation of network protocols. Autom. Control Comput. Sci. **49**(8), 826–833 (2015)
8. Zegzhda, P., Zegzhda, D., Pavlenko, E., Dremov, A.: Detecting android application malicious behaviors based on the analysis of control flows and data flows. In: Proceedings of the 10th International Conference on Security of Information and Networks, pp. 280–283 (2017)
9. Abramov, G., Korobova, L., Ivashin, A., Matytsina, I.: Information system for diagnosis of respiratory system diseases. In: Journal of Physics: Conference Series, vol. 1015, p. 042036. IOP Publishing, Bristol (2018)
10. Barabanov, A.V., Markov, A.S., Tsirlov, V.L.: Methodological framework for analysis and synthesis of a set of secure software development controls. J. Theor. Appl. Inf. Technol. **88**(1), 77–88(2016)
11. Howard, M., Lipner, S.: The Security Development Lifecycle: A Process for Developing Demonstrably More Secure Software. Microsoft Press, Redmond (2006)
12. Calzavara, S., Focardi, R., Nemec, M., Rabitti, A., Squarcina, M.: Postcards from the post-http world: amplification of https vulnerabilities in the web ecosystem. In: 2019 IEEE Symposium on Security and Privacy (SP), pp. 281–298. IEEE, Piscataway (2019)
13. Calzavara, S., Focardi, R., Squarcina, M., Tempesta, M.: Surviving the web: a journey into web session security. ACM Comput. Surv. **50**(1), 1–34 (2017)
14. Nirmal, K., Janet, B., Kumar, R.: Web application vulnerabilities-the hacker's treasure. In: 2018 International Conference on Inventive Research in Computing Applications (ICIRCA), pp. 58–62. IEEE, Piscataway (2018)

15. Petrenko, A.S., Petrenko, S.A., Makoveichuk, K.A., Chetyrbok, P.V.: Protection model of PCS of subway from attacks type wanna cry, petya and bad rabbit IoT. In: 2018 IEEE Conference of Russian Young Researchers in Electrical and Electronic Engineering (EIConRus), pp. 945–949. IEEE, Piscataway (2018)
16. Priya, R., Lifna, C., Jagli, D., Joy, A.: Rational unified treatment for web application vulnerability assessment. In: 2014 International Conference on Circuits, Systems, Communication and Information Technology Applications (CSCITA), pp. 336–340. IEEE, Piscataway (2014)
17. Bhor, R., Khanuja, H.: Analysis of web application security mechanism and attack detection using vulnerability injection technique. In: 2016 International Conference on Computing Communication Control and Automation (ICCUBEA), pp. 1–6. IEEE, Piscataway (2016)
18. Wang, B., Liu, L., Li, F., Zhang, J., Chen, T., Zou, Z.: Research on web application security vulnerability scanning technology. In: 2019 IEEE 4th Advanced Information Technology, Electronic and Automation Control Conference (IAEAC), vol. 1, pp. 1524–1528. IEEE, Piscataway (2019)
19. Yadav, D., Gupta, D., Singh, D., Kumar, D., Sharma, U.: Vulnerabilities and security of web applications. In: 2018 4th International Conference on Computing Communication and Automation (ICCCA), pp. 1–5. IEEE, Piscataway (2018)
20. Malkawi, M., Özyer, T., Alhajj, R.: Automation of active reconnaissancephase: an automated api-based port and vulnerability scanner. In: Proceedings of the 2021 IEEE/ACM International Conference on Advances in Social Networks Analysis and Mining. ASONAM'21, pp. 622–629. Association for Computing Machinery, New York (2021). https://doi.org/10.1145/3487351.3492720
21. Varenitca, V.V., Markov, A.S., Savchenko, V.V.: Recommended practices for the analysis of web application vulnerabilities. In: 10th Anniversary International Scientific and Technical Conference on Secure Information Technologies, BIT 2019 CEUR Workshop Proceedings, vol. 2603, pp. 75–78 (2019)
22. Marshmallow: Simplified Object Serialization — Marshmallow 3.15.0 Documentation. https://marshmallow.readthedocs.io/en/stable/
23. Shah, M., Ahmed, S., Saeed, K., Junaid, M., Khan, H., et al.: Penetration testing active reconnaissance phase–optimized port scanning with nmap tool. In: 2019 2nd International Conference on Computing, Mathematics and Engineering Technologies (iCoMET), pp. 1–6. IEEE, Piscataway (2019)
24. Chakrabarti, S., Chakraborty, M., Mukhopadhyay, I.: Study of snortbased ids. In: Proceedings of the International Conference and Workshop on Emerging Trends in Technology, pp. 43–47 (2010)
25. Kaur, G., Kaur, N.: Penetration testing–reconnaissance with NMAP tool. Int. J. Adv. Res. Comput. Sci. **8**(3), 844–846 (2017)
26. Durumeric, Z., Wustrow, E., Halderman, J.A.: ZMap: fast internet-wide scanning and its security applications. In: 22nd {USENIX} Security Symposium ({USENIX} Security'13), pp. 605–620 (2013)
27. Schagen, N., Koning, K., Bos, H., Giuffrida, C.: Towards automated vulnerability scanning of network servers. In: Proceedings of the 11th European Workshop on Systems Security, pp. 1–6 (2018)
28. Roy, A., Mejia, L., Helling, P., Olmsted, A.: Automation of cyberreconnaissance: a java-based open source tool for information gathering. In: 2017 12th International Conference for Internet Technology and Secured Transactions (ICITST), pp. 424–426. IEEE, Piscataway (2017)
29. Panjwani, S., Tan, S., Jarrin, K.M., Cukier, M.: An experimental evaluation to determine if port scans are precursors to an attack. In: 2005 International Conference on Dependable Systems and Networks (DSN'05), pp. 602–611. IEEE, Piscataway (2005)
30. Zhao, J.J., Zhao, S.Y.: Opportunities and threats: a security assessmentof state e-government websites. Govt. Inf. Quart. **27**(1), 49–56 (2010)
31. Mooers, C.N.: Preventing software piracy. Computer **10**(3), 29–30 (1977)

32. McPherson, J., Ma, K.-L., Krystosk, P., Bartoletti, T., Christensen, M.: Portvis: a tool for port-based detection of security events. In: Proceedings of the 2004 ACM Workshop on Visualization and Data Mining for Computer Security, pp. 73–81 (2004)
33. Mathew, K., Tabassum, M., Siok, M.V.L.A.: A study of open ports as security vulnerabilities in common user computers. In: 2014 International Conference on Computational Science and Technology (ICCST), pp. 1–6. IEEE, Piscataway (2014)
34. Maini, R., Bvducoep, P., Pandey, R., Kumar, R., Gupta, R.: Automated web vulnerability scanner. Int. J. Eng. Appl. Sci. Technol. **4**(1), 132–136 (2019)
35. What Is Python? Executive Summary Python.org. https://www.python.org/doc/essays/blurb/
36. Van Rossum, G., et al.: Python programming language. In: USENIX Annual Technical Conference, vol. 41, pp. 1–36 (2007)
37. Orebaugh, A., Pinkard, B.: Nmap in the Enterprise: Your Guide to Network Scanning. Elsevier, Amsterdam (2011)
38. Lyon, G.F.: Nmap Network Scanning: The Official Nmap Project Guide to Network Discovery and Security Scanning. Insecure. Com LLC (US) (2008)
39. Liao, S., Zhou, C., Zhao, Y., Zhang, Z., Zhang, C., Gao, Y., Zhong, G.: A comprehensive detection approach of nmap: principles, rules and experiments. In: 2020 International Conference on Cyber-Enabled Distributed Computing and Knowledge Discovery (CyberC), pp. 64–71. IEEE, Piscataway (2020)
40. Chapter 15. Nmap Reference Guide `Nmap Network Scanning`. https://nmap.org/book/man.html#man-description
41. Nmap Scripting Engine (NSE) `Nmap Network Scanning`. https://nmap.org/book/man.html#man-description
42. Grinberg, M.: Flask Web Development: Developing Web Applications with Python. O'Reilly Media, Sebastopol (2018)
43. Masse, M.: REST API Design Rulebook: Designing Consistent RESTful Web Service Interfaces. O'Reilly Media, Sebastopol (2011)
44. Burr, C., Couturier, B.: A gateway between gitlab ci and dirac. In: EPJ Web of Conferences, vol. 245, p. 05026. EDP Sciences, Les Ulis (2020)
45. Ionescu, V.M.: The analysis of the performance of rabbitmq and activemq. In: 2015 14th RoEduNet International Conference-Networking in Education and Research (RoEduNet NER), pp. 132–137. IEEE, Piscataway (2015)
46. What Can RabbitMQ do for You? — RabbitMQ. https://www.rabbitmq.com/features.html
47. RabbitMQ Tutorial - "Hello World!" — RabbitMQ. https://www.rabbitmq.com/tutorials/tutorial-one-python.html
48. Flower - Celery Monitoring Tool – Flower 1.0.1 Documentation. https://flower.readthedocs.io/en/latest/
49. Castiglione, A., Palmieri, F., Petraglia, M., Pizzolante, R.: Vulsploit: A module for semi-automatic exploitation of vulnerabilities. In: IFIP International Conference on Testing Software and Systems, pp. 89–103. Springer, Berlin (2020)

Using Smart Glasses for Monitoring Cyber Threat Intelligence Feeds in a Multitasking Environment

Mikko Korkiakoski, Febrian Setianto, Fatima Sadiq, Ummi Khaira Latif, Paula Alavesa, and Panos Kostakos

Abstract The surge of COVID-19 has introduced a new threat surface as malevolent actors are trying to benefit from the pandemic. Because of this, new information sources and visualization tools about COVID-19 have been introduced into the workflow of frontline practitioners. As a result, analysts are increasingly required to shift their focus between different visual displays to monitor pandemic-related data, security threats, and incidents. Augmented reality (AR) smart glasses can overlay digital data to the physical environment in a comprehensible manner. However, the real-life use situations are often complex and require fast knowledge acquisition from multiple sources. In this study, we report results from a pilot and a trial where we used an overlaid AR information interface coupled with traditional computer monitors. Both the pilot and the study had six test subjects. Smart glasses and their applications often have a limited amount of adjustments for individual differences in physiology, which may influence task performance in fast-paced situations. Our goal was to evaluate a multi-tasking setup with traditional monitors and an AR headset where notifications from the new COVID-19 Malware Information Sharing Platform (MISP) instance were visualized along with some interactability. Our results indicate some gender differences in some aspects of situational awareness. Some parts of the task load index also showed correlation with certain aspects of situational awareness. While our experiments were small scale, in this study, we showed that AR has the potential to improve situational awareness and task performance in monitoring tasks.

Keywords Augmented reality · Smart glasses · HoloLens 2 · MISP · Immersive analytics · COVID-19

M. Korkiakoski (✉) · F. Setianto · F. Sadiq · U. K. Latif · P. Alavesa · P. Kostakos
University of Oulu, Center for Ubiquitous Computing, Oulu, Finland
e-mail: Mikko.Korkiakoski@oulu.fi

S. T. Özyer, B. Kaya (eds.), *Cyber Security and Social Media Applications*, Lecture Notes in Social Networks, https://doi.org/10.1007/978-3-031-33065-0_3

1 Introduction

The outbreak of COVID-19 has introduced mandatory lockdown policies and strict social distancing protocols that have majorly interrupted social-economic life in most countries. Along with the health crisis, we have also seen a surging information crisis where large volumes of information across multiple domains urgently need to be shared, analyzed, authenticated, and verified. Against this backdrop, many pre-existing cybersecurity challenges have been exacerbated and often morphed into complex hybrid phenomena. In other words, many of the threats we are now facing are not new, but new tools are needed in fighting those disruptions [1–3].

Past research has shown that smart glasses have an impact on comprehension [4]. Augmented reality (AR) allows embedding semantically familiar material such as objects to an environment while retaining spatial information. This can enhance learning experiences as it complements perception. However, the real benefit of AR and any cross-reality lies in the possibility to visualize abstract digital data with spatial information, to aid in both comprehension and situational awareness (SA) [5, 6]. While the aerospace and defense sectors are utilizing the technology, the adoption of AR by the cybersecurity community is still lagging. In previous work, the authors presented a pilot study that gauged the SA of six participants who assumed the role of a cybersecurity analyst facilitated by a custom immersive AR analytics tool [7]. Preliminary results have indicated a gender gap that warrants further investigation. We performed a follow-up study with additional subjects and implemented new features that focused on degendering the experimental set-up.

Given the lack of research in this area, the current experiment was driven by the following research questions: Does the consolidation of cyber threat intelligence feeds into a single display improve information comprehension? Is there a gender gap that might affect the performance of the analyst exposed to AR tools? This article is structured as follows. First, we introduce the technologies, tools, and concepts our prototype applications are based on. Next, the applications, their development, and the study procedure is detailed, beginning in each section with the pilot description followed by the experiment. Then we present the results from our pilot and the subsequent experiment. The discussion section contemplates the relevance of our results and talks about future work and limitations. The article ends with conclusions.

2 Applications and Tools

2.1 TIPs and the MISP

With the increased usage of the internet, the type and number of cyber attacks has also increased. This spawns from the fact that originally the main purpose of the internet was connectivity and not security. Over time, the nature of these cyber threats has changed in terms of scope, impact, and complexity. This change has also created an urgent need to develop information-sharing platforms to share threats and mitigation strategies with each other, in order to help protect porous infrastructures against attacks. In response to this, new tools have been introduced by the software industry known as Threat Intelligence Platforms (TIP), enabling organizations to share threat information with the community and help mitigate risks. Threat intelligence in simple terms can be described as the knowledge and information of threats, their capabilities, infrastructure, motives, goals, and resources. TIPs are the tools that help in getting the threat intelligence information into the form of security data, and then analyzing it to identify and mitigate security threats [8].

TIPs can be classified into two major categories based on their availability: Open source (free) and commercial (paid). Alien Vault open threat intelligence (OTX) and Malware Information Sharing Platform (MISP) are free to use, while commercial TIPs include the Accenture cyber threat intelligence platform, Anomali ThreatStream, IBM-Xforce Exchange, etc. These platforms have different agendas and goals such as providing threat sharing, sharing aggregated data, indicators of compromise, and supporting different standards and licensing models [9].

Currently, the most highlighted and widely used threat intelligence platform is the MISP. It is a community and government-driven open source trusted platform developed by The Computer Incident Response Center Luxembourg (CIRCL). In MISP, information usually refers to Indicators of Compromise (IOC) and information about cybersecurity threats and incidents. Furthermore, MISP provides information about vulnerabilities and financial indicators used in fraud cases. An entry in MISP is known as an event, containing the information, IOCs, and attachments in the form of attributes (category and type). MISP has several sharing models such as sharing information with organization only, community only, with some connected communities, or with all MISP communities. MISP has its own taxonomy for sharing information and tagging them correctly. It has a REST client and an API for data pulling and pushing purposes. MISP has many instances, events, and attributes and is widely used for data extraction and research [10].

2.2 *The COVID-19 MISP Instance*

Since December 2019, the world has been struggling with a COVID-19 pandemic. The effect of COVID-19 on the entire world has been immense and it has also affected every sector of the world's economy. Many organizations and researchers took advantage of this time and tried to explore and track COVID-19 data in real-time by visualizing them using dashboards. These dashboards make it easy for the layman to see and understand the day-to-day numbers. But over time, cyber criminals also moved into the world of COVID-19 and began launching attacks, scams, and spreading misinformation and fake data. To mitigate any risks, CIRCL

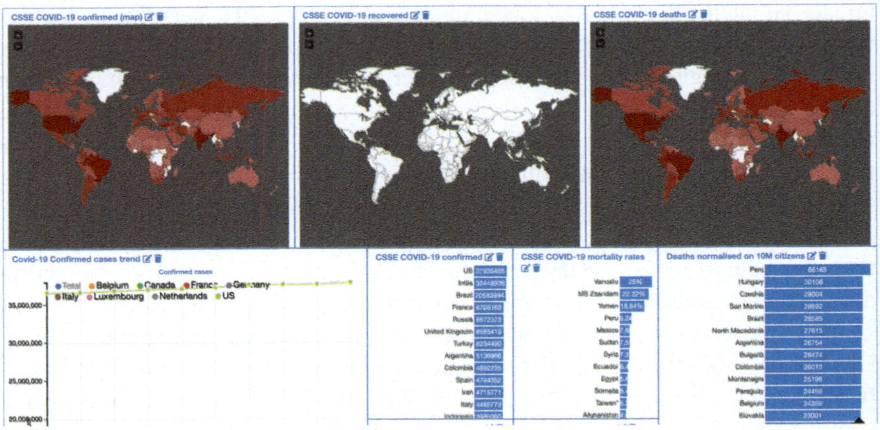

Fig. 1 MISP Covid-19 dashboard

introduced a new retrofitted instance of MISP that contains feeds and events related to COVID-19 [11]. The instance also includes a custom dashboard with official statistics for COVID-19 deaths, confirmed cases, and recovered cases by country (Fig. 1).

The events home page is the same as the MISP main instance but with COVID-19-based event content. On this page, users can find all the events with fake/abusive content related to COVID-19. These events include misinformation, fake data, fake COVID-19 domains, and breaching alerts. It is also possible to extract data from this instance either by using a cURL command or by using a Python library called PyMISP that enables access to MISP platforms via their REST API. For both cURL and PyMISP, it is mandatory to mention the MISP instance URL parameters and the unique key of the user to download the data either in JSON or comma-separated values (CSV) format [12]. The extracted data includes events and their attributes (Fig. 2). The attributes include a unique id, event information, analysis, threat level id, etc.

2.3 Immersive Security Analytics

Immersive Analytics (IA) is a method of using immersion to better understand and analyze complex data and to aid in decision making [13]. Immersive Security Analytics (ISA) is a sub-category of IA that specifically concentrates in the immersive analytics of cybersecurity data. The type of system immersion in question [14], can be achieved through various different methods including AR, Virtual Reality (VR), Extended Reality (XR), Virtual Reality Environments (VRE), touch displays, or a combination of these methods.

Event Info
Threat Level 1
"Coronavirus Finally Hit us Bewarned !!!" - Coronavirus-themed Phishing Lure Delivering Agent Tesla
(U) Spearphishing Attack Uses COVID-21 Lure to Target Ukrainian Government
CovidLock Update: Deeper Analysis of Coronavirus Android Ransomware
Threat Level 2
Operation Poisoned News: Hong Kong Users Targeted With Mobile Malware via Local News Links
Threat Level 3
"Subject: 2nd-COVID Outbreak Warning Alerts For Business Owners/Workers Globally !!! - From: kerwin.john@slaspa.com" Campaign
"Subject: Cigo covid inps circolare 115 2020 - From: l.migliorini@confindustriatoscanasud.it" Campaign
Criminals in a festive mood
OSINT - New Mac backdoor using antiquated code
OSINT - Death Comes Calling: Thanatos/Alphabot Trojan Hits the Market
Threat Level 4
AlienVault | Cyber Attacks Leveraging the COVID-19/CoronaVirus Pandemic
AlienVault | Cybercriminals Prey on the COVID-19 Pandemic
AlienVault | Fake Corona Antivirus distributes BlackNET remote administration tool
AlienVault | Fake COVID-19 survey hides ransomware in Canadian university attack
Covid19 Stats Pakistan: Total 534, Deaths 3, Recovered 5
Covid19Malware feed
Cyber_Hat | APT29 targets COVID-19 vaccine development
Cyber_Hat | APT36 jumps on the coronavirus bandwagon, delivers CrimsonRAT
Cyber_Hat | attack activities using COVID
Grand Total

Fig. 2 Example events for different threat levels in MISP

IA solutions are now increasingly adapted to improve data analysis and visualization, but cybersecurity tools are still mostly using older, classical graphical representations (2D dashboards), and are significantly lagging behind utilizing the full potential of the previously mentioned solutions [15]. So, to make ISA a valid option, it should incorporate some types of intuitive interactions in order for the users of such systems to be able to concentrate on doing the actual work and not think about how to perform simple actions. Intuitive interactions in principle means: to make functions, locations, and appearance familiar for features that are already known, making the usage of less well-known features feel more obvious by using familiar things to demonstrate their function/appearance/location, and increasing the consistency within the interface so that function/appearance/location features are consistent in all parts of the design, etc. [16]

While I(S)A offers new ways of analyzing and observing data, it does come with its own challenges. When monitoring a system using AR, the view of the user can quickly become cluttered with all sorts of data and this can lead to an information overload, just like when using multiple regular displays. The solution here could be to share the information between regular displays and AR and use a drill-down style so the user can get more information quickly and easily when needed.

Another challenge is the complexity of the used ISA system. Operating and interacting with a system that groups traditional displays, head mounted displays (HMD), tablets, controllers, etc. into one large entity, can easily become complicated and taxing to the user. Another issue related to this is what is being called as having "gorilla arms". What this means, is when a user is constantly using their arms or only just keeping them up, they quickly get tired. This problem could perhaps be helped by using voice controls or even gaze/eye controls [17].

Not just the design or software solution cause issues with individuals while observing data visualization in AR or VR. The size dimensions of the headsets and their straps including the displays' standard or allowed range for adjusting the inter-pupillary distance (IPD) become an issue when the users' physiology is outside the optimal range [18]. In many XR headsets, like Quest 2, there is a software correction for IPD, however, this correction narrows the field of view (FOV) on the highest IPD setting [19]. While VR HMDs do not have the same issues with lenses as the AR HMD, the IPD is still an issue. HoloLens 2 has a FOV of 30°, however, this FOV is situated on the displays according to IPD adjustment, which has again a limited scale. Gender difference in mean values of IPDs among white American subjects is not substantial, only 2–3 mm [20], but on an AR HMD that is near the pupils, this difference may influence the experience. Slight differences especially in experiencing and task performance between different gendered participants using VR HMDs have been observed, but with very mixed results. However, findings on cognition and memory are heterogeneous [21]. There are recent studies on gender differences using HoloLens 1 but HoloLens 2 is substantially better in terms of FOV, IPD adjustment, and display resolution, so the findings between technologies may not be comparable.

2.4 Situational Awareness with Smart Glasses

VR headsets are extensively used for immersive training and simulating extreme scenarios [22]. However, while smart glasses have been initially used as simple front-end displays, new devices have much better computing power and are able to perform more complex tasks. And as the technology is maturing, different types of AR devices and applications are becoming more mainstream. AR can already be effectively used to help in tasks like vehicle operating [23, 24], system monitoring [25], drone operating [26], military field operations [27] etc. The AR implementations most commonly used in different tasks involve projecting information or objects onto a perceived real-world environment to assist and inform the user. More importantly, AR can be used to help keep the user's focus on the important parts of the system. This is usually accomplished by gathering and projecting system critical information, traditionally scattered in various places and/or multiple monitors, into the AR view. This approach frees the user from having to monitor multiple places for critical information, and they can more effectively focus on their main tasks. AR implementations are often used as tools to help the operator focus on a main task, rather than making AR itself the main focus. Pascale et al. [25] experimented with using AR to keep nurses up to date on patients' vital signs in order to help them better focus on a main task of doing dosage calculations. This is similar to our experiment, where there is a main task that is done using a regular display, pen and paper, while monitoring other information via AR overlays. This kind of multitasking helper function spreads across many fields. For example, Rowen et al. [28] and Hong et al. [24] implemented AR in a similar fashion in order to help boat operators focus on their main task of navigation while Coleman et al. [26] used a similar method to help keep the focus of drone operators on the drone's video feed.

When it comes to SA, using an AR application can have a positive effect but it can also increase the attention demand as found in two studies on telemaintenance and teleoperation [5, 29]. In these experiments, a VR application actually performed better in terms of SA and while the AR implementation improved SA it also increased attention demand more than the VR application. In many situations though, VR and AR both are not a feasible option. For vehicle operating and system monitoring AR is the far superior, and in most cases, the only realistic option. It has been shown that using AR for guidance, safety, and general information affects SA in a positive way in a maritime environment [24, 28] and driving a car on public roads [23]. Similarly, a positive SA effect was recorded in an experiment by Pascale et al. [25], where a nurse was monitoring the vital information of several patients via AR while at the same time performing another task outside the AR space.

3 Design and Implementation

The user study was conducted in two parts: first, a pilot study which was followed by a more robust user study. The main improvement between the studies was in the setup, however, the application was also improved. In the following, the pilot application is described first, followed by the description of the study application.

3.1 *Pilot: Design and Setup*

In the pilot experiment, we tested an AR overlaid information interface coupled with a regular dual-monitor desktop setup. Testing was conducted using a within-subjects design with one condition, so each participant tested the same system and did the same tasks. The experiment simulates the work space of a cybersecurity analyst/monitor. The setup features a desktop computer with dual monitors, one of which is used for finding information and the other for experiment-related recording etc. The AR device (HoloLens 2) overlays a threat level indicator in between the two monitors and an event information panel next to the left side monitor. This setup simulates multi-tasking between a main working task using the keyboard, mouse, and the right side monitor and a secondary task of monitoring cybersecurity information presented on the left side panel. Participants were not able to make any adjustments to the location and height of either the virtual or the physical work station.

For the later study, the basic layout stayed the same but the location was changed. The interface system was also improved and new functionality was added to the AR system.

3.2 Pilot: Technical Implementation

The HoloLens 2 application was created with Unity game engine using Mixed Reality Toolkit and C# scripts. For connecting to the COVID-19 database in MISP, UnityWebRequest was used. The MISP instance exposes an API which in return gives the caller a JSON-structured response. The data received via the UnityWebRequest is temporarily stored in the unit's memory. First, the data is parsed with SimpleJSON [30] to support exploration and iteration. Next, a custom algorithm is used to randomly pick 60 events, 15 for each of the 4 threat levels. The event information is then displayed on an AR projected panel situated roughly 45° left of the user. Because the COVID-19 database consists mostly of level 4 threats, the system shows a maximum of 5 events for each threat level, until moving onto a new cycle. This cycle is repeated 3 times for a total of 60 events (15 events × 4 threat levels). The event information panel displays each event for roughly 5 s. The threat level of the event is also visualized using a color-coded floating sphere that is projected in front of the user's viewpoint. The colors relate to the threat levels as follows: white = threat level 4; green = threat level 3; yellow = threat level 2; and red = threat level 1. Level 1 is the highest threat level and level 4 the lowest. The color of the ball and the event information are updated at the same time, so the user will know the threat level of the current event, just by glancing at the color of the object (Fig. 3).

Fig. 3 Pilot experiment setup with the threat level indicator in the middle (green sphere) and the event information panel on the left

3.3 Study: Design and Setup

For the study, the development continued with Unity and HoloLens 2. Numerous changes were made to the original design. The information visible to the user by default was minimized to show only the relevant information of event threat level and the time the event came in. The event threat level was again also indicated by the color of the sphere. Although the default view was minimal, the user had the ability to get relevant information about the latest event as well as events related to that particular event. An option to show a brief event history (3. in Fig. 4) was also added but the use of this was not mandatory to complete the tasks given in this study. The last addition information-wise was the MISP dashboard. The user could now start up the MISP dashboard inside the AR system and find visualizations and statistics as shown in Figs. 1 and 5.

In the pilot, the setup of the system was static in the sense that once the system was started, the user could not change anything or interact with any parts of the AR system. For the study, this was one area that saw major improvements. Gesture controls were added along with voice commands, interactable buttons, and new information panels. Now the user could use what is called an "air tap" to press buttons and panels from afar. Naturally, all buttons could also be pressed just like in real life. The user could launch a MISP dashboard instance in a browser and a related events panel by using voice commands. The MISP dashboard and the related events panel also had a dedicated button for each of them to be used instead if the user did not want to use the voice commands. The event history log was also accessible via a button.

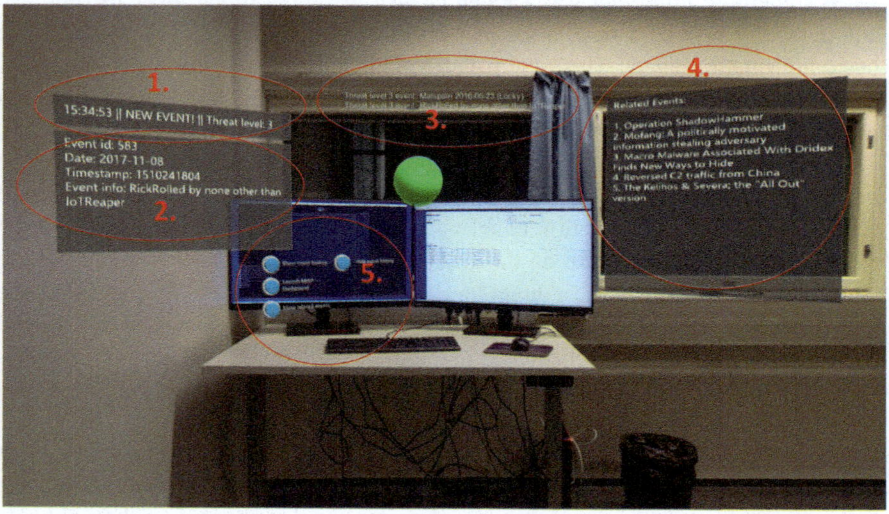

Fig. 4 Improved design/layout: 1. New event panel, 2. Additional information panel, 3. Event log, 4. Related events panel, 5. Interactable buttons

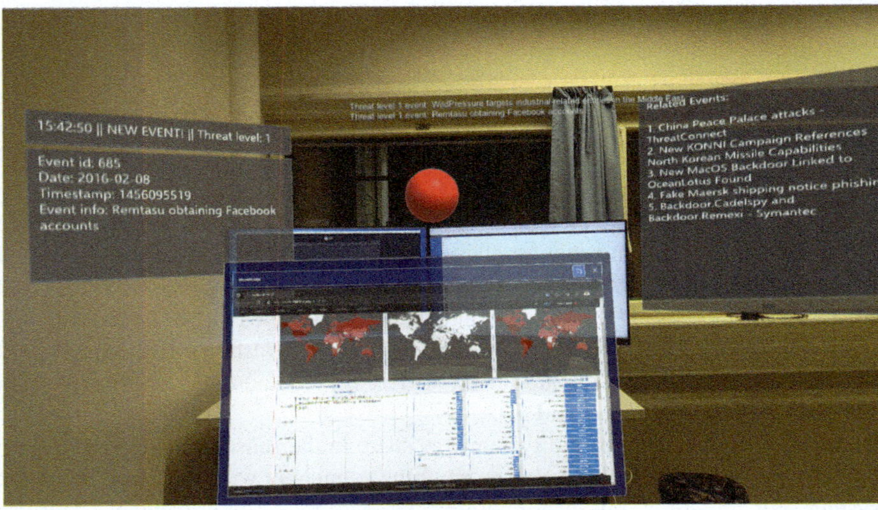

Fig. 5 Improved layout with MISP dashboard opened

In the new layout, only the placement of the sphere and what used to be the information panel was kept the same as in the pilot. As stated before, the old pilot information panel was minimized to show only the threat level and time of arrival (1. in Fig. 4). The sphere also worked as in the pilot, by changing its color according to the threat level of the newest event. But this time an alert sound was added for level 1 threats to create a multi-sensory alert system to notify the user when something important was happening. An incoming threat level 1 event also automatically opened a new additional information panel (2. in Fig. 4) under the basic information panel. This new panel would show the relevant information for the event. This functionality was added to enable quick access to crucial information while still keeping the system interface as minimal as possible at all other times. The additional information panel would automatically close when the level 1 treat passed and a new event with lesser importance came in. However, the user could open the additional information panel by using the aforementioned air tap on the information panel. An air tap on the additional information panel would close the panel at any time, even while a threat level 1 event was displayed. When needed, the user could access a related events panel (4. in Fig. 4) to see 5 previous events related to the event currently being displayed in the information panel. This panel could be opened at any point via its button or a voice command. The panel would be closed by air tapping on it.

The new system layout can be seen in Fig. 4. The event sphere and the new event panel are situated as in the pilot. The additional information panel opens up just below the new event panel and the related events panel is situated on the right side of the user, at the same angle and relative position as the panels on the left side. Buttons are tilted towards the user and within arm's reach just to the left of the center position (5. in Fig. 4). The MISP dashboard opens up slightly below the user's viewpoint (Fig. 5) and can be freely moved and resized when needed. In addition to these layout changes, the event picking algorithm was also slightly altered. Instead of showing 5 events per threat level for 3 cycles with a minimum hold time of 5 s minimum, the new algorithm shows 3 events per threat level for 4 cycles with an event hold time of 10 s minimum. This change reduced the total amount of events from 60 to 48 but increased the minimum experiment time from 6 min to 8 min. On average the overall experiment time changed from roughly 7 min in the pilot to 9 min in the study. The algorithm sometimes needs more than the minimum time to find the correct type of an event so there is always some slight variances in the overall experiment duration. The variation though is not more than 5–10 s between the fastest and slowest experiment time. The algorithm was changed to give the users more time to properly log the IDs for alerts, observe the related events, and to give them more time in general due to the increased number of tasks.

4 Experimental Sessions

In the following, the experimental procedure of the pilot is described first, followed by the study procedure description.

4.1 Pilot: Participants

Given the COVID-19 limitations, we recruited a sample of convenience from the student population. For the pilot the participant pool comprised of 6 people in total (3 male and 3 female). Both the initial and the follow-up experiments were conducted at the Center for Ubiquitous Computing, University of Oulu, Finland.

4.2 Pilot: Experimental Procedure

Before the explanation of the experiment, the participants signed a consent form and filled in their age and gender. The participants were told that they would assume the identity of a cybersecurity expert monitoring COVID-19-related cybersecurity information using HoloLens 2 AR smart glasses while at the same time completing a separate main task. In the initial experiment, the main task was to find answers to 10 pre-determined questions, using a provided web browser (right side monitor in Fig. 3). The participants were also told that they could take notes about the constantly updating security events (for example write down event id numbers and corresponding threat level), but that it was up to them to decide whether to do this secondary task or not. Next, the AR implementation and the information it displayed were explained. The participants' desktop arrangement including the AR projected overlays can be seen in Fig. 3. Lastly, the participants were informed that after the experiment, there would be a short questionnaire.

4.3 Pilot: Materials

After the experiment was started, the 10 questions were revealed to the participants. They were also provided with a sheet of paper for answering the questions which also included a table to write down event IDs and their corresponding threat levels if they chose to do so. The researcher silently followed the experiment and did not talk to the participants unless they asked a question relating to the experiment procedure or if they needed help with something (for example got stuck in searching for the answer to the first question). The experiment would last roughly 7 min, the time varying a bit depending on the accuracy of the randomized event finding algorithm (see next section). After the test was finished, the participants filled out a Situational Awareness Rating Technique (SART) questionnaire. Video recordings were taken from 4 out of 6 experiments for verification purposes. The videos were recorded via HoloLens 2's integrated camera so the recordings would accurately show where the participants' view was focused during the experiment. Before and after each experiment, the HoloLens 2 was sanitized using Cleanbox, a machine that uses ultraviolet light to kill bacteria, made specifically for disinfecting masks, eye-wear, and headsets.

4.4 Study: Participants

As with the pilot, we conducted the study with six participants (3 male and 3 female). The sizes of the gender groups stayed the same but three participants changed from the pilot (1 male and 2 female).

4.5 Study: Experiment Procedure

To make the instructions and testing procedure more uniform, a number of changes were made to the protocol after the pilot. As before, the participants first signed a consent form but this time all the instructions regarding the testing procedure were written down. This way all the participants received the same information about how the system worked, the tasks, system controls, and some general information about the test. After they had read the instructions, the researcher would launch the app and show the participants how everything looked and worked in the AR system (see Fig. 4 for new design). This was done using a HoloLens 2 feature, where the view of the user can be broadcast as a live preview on a PC monitor. During the demonstration and afterwards the participants were asked if they had any questions about the functionality of the system, their tasks, or if they simply had questions about anything that was unclear. After the demonstration was over, the HoloLens 2 unit was disinfected and given to the participant. When the participant felt confident about the experiment they would first adjust the table at a height they wanted and then launched the application themselves. It was at this point that the experiment started and the 10 questions were revealed just like in the pilot.

All participants were also notified beforehand that they should not speak to the researcher or ask any questions unless there was an error in the system or if they had problems doing their tasks. This rule was implemented to reduce bias towards participants of the opposite sex. Only two participants were talked to during the experiment (one male and one female). One of them was trying to use the MISP dashboard browser to look for the answers for task 1, so they were quickly reminded to use the regular display/mouse/keyboard to do the questions part. The other participant was reminded about the questions because from the beginning it seemed to the researcher like the user had forgotten the existence of the questions altogether. This time there were four tasks to be completed. Tasks 1 and 2 were of equal importance (high) and task 3 and 4 were less important additional tasks. This was made clear to the participants prior to the experiment.

The tasks were as follows:

1. Answer the provided 10 questions
2. Log all threat level 1 event IDs
3. Log one related event for an event with a threat level of 2, 3 and 4
4. Find the country with the 2nd most COVID-19-related deaths normalized on 10M people and the number of deaths for it.

The experiment would end when the algorithm had cycled through all the 48 events and then prompt back to the starting state and informing the user with an alarm and a text in the event panel saying: "Test finished, bye".

4.6 Study: Materials

To evaluate the SA of users during the experiments the SART [31], which is a post-trial subjective rating technique, was administered to all participants [32]. SART was used in both the pilot experiment and the succeeding study. The questionnaire was kept the same in both cases so the results could possibly be compared between the two. The SART score is calculated with the formulas

$$SA = U - (D - S)$$

and here specifically:

$$\sum(Q8 + Q9) - \left(\sum(Q1 + Q2 + Q3) - \sum(Q4 + Q5 + Q6 + Q7)\right)$$

where U is the summed understanding, D the summed demand, and S is the summed supply (limits: −14 and 46). The questions used for SART are as follows:

1. **Instability of the situation**: How changeable was the situation? Is the situation highly unstable and likely to change suddenly (high) or is it very stable and straightforward (low)?
2. **Complexity of the situation**: How complicated is the situation? Is it complex with many interrelated components (high) or is it simple and straightforward (low)?
3. **Variability of the situation**: How many variables are changing within the situation? Are there a large number of factors varying (high) or are the very few variables changing (low)?
4. **Arousal**: How aroused are you in the situation? Are you alert and ready for activity (high) or do you have a low degree of alertness (low)?
5. **Concentration of attention**: How much are you concentrating on the situation? Are you concentrating on many aspects of the situation (high) or focused on only one (low)?
6. **Division of attention**: How much is your attention divided in the situation? Are you concentrating on many aspects of the situation (high) or focused on only one (low)?
7. **Spare mental capacity**: How much mental capacity do you have to spare in the situation? Do you have sufficient mental capacity to attend to many variables (high) or nothing to spare at all (low)?
8. **Information quantity**: How much information have you gained about the situation? Have you received and understood a great deal of knowledge (high) or very little (low)?
9. **Familiarity with situation**: How familiar are you with the situation? Do you have a great deal of relevant experience (high) or is it a new situation (low)?

To assess the workload of the users during the experiment, the National Aeronautics and Space Administration Task Load Index (NASA TLX) was used. NASA TLX was only used in the study, not the pilot. It was introduced as an additional tool because the study had an increased number of tasks and features and it would be interesting to see if the workload had any correlation with situational awareness or task performance. NASA TLX uses 21-point scales to measure the task load of a user in six different categories [33]. The questionnaire categories are as follows:

1. **Mental Demand**—How mentally demanding was the task?
2. **Physical Demand**—How physically demanding was the task?
3. **Temporal Demand**—How hurried or rushed was the pace of the task?
4. **Performance**—How successful were you in accomplishing what you were asked to do?

5. **Effort**—How hard did you have to work to accomplish your level of performance?
6. **Frustration**—How insecure, discouraged, irritated, stressed and annoyed were you?

After evaluating their task load in the six categories the user was presented with 15 different pairings formed from the same six categories. From each of the pairs (Fig. 6), the user would choose the one they thought was the more important factor in contributing to the overall workload. From these answers, we can determine the weight for each of the categories and from there the adjusted ratings for each category can be calculated [33].

For example, we get the adjusted rating for Mental Demand as follows:

1. The user reports Mental Demand as 13 (raw rating)

Effort Or Performance	Temporal Demand Or Frustration	Temporal Demand Or Effort	Physical Demand Or Frustration
Performance Or Frustration	Physical Demand Or Temporal Demand	Physical Demand Or Performance	Temporal Demand Or Mental Demand
Frustration Or Effort	Performance Or Mental Demand	Performance Or Temporal Demand	Mental Demand Or Effort
Mental Demand Or Physical Demand	Effort Or Physical Demand	Frustration Or Mental Demand	

Fig. 6 NASA TLX pairing groups. From each of these pairs the participants choose the one that they think was a more contributing factor to the overall workload. These answers are then used in the workload calculations.

2. The user picks Mental Demand as the more important factor in 4 of the pairings (weight)

Then by using this formula we can calculate the adjusted rating for Mental Demand:

$$AdjustedRating = RawRating \times Weight$$

$$AdjustedRating = 13 \times 4 = 52$$

The same procedure is repeated for all 6 categories. The total adjusted workload is calculated simply by adding up the adjusted ratings of all categories. The weighted rating of this adjusted workload can be obtained by dividing it by the number of pairings used to determine the weights (15):

$$\sum(AdjustedRatings)/15$$

5 Results

In this results section we compare the results between users, but also results between genders. When we talk about the participants with more experience, it specifically means that these users participated in the pilot as well as the study. So these participants already had experience with the base AR layout as well as how the main tasks were performed.

5.1 Pilot

The gender differences were analyzed using one-way analysis of variance (ANOVA) and the correlations between SART scores and task performance with Pearson's correlation test (p-value for both tests was chosen as 0.05). We are mostly interested in the SART scores as well as the performance in the 2 tasks. The final SART scores did not show any significant differences between the genders Table 1. But when looking at the individual components of the scores, some differences do stand out. More specifically there were significant differences between males and females in questions 5 ($p = 0.033$) and 6 ($p = 0.005$). For question 5: *"How much are you concentrating on the situation? Are you concentrating on many aspects of the situation or focused on only one?"* and question 6: *"How much is your attention divided in the situation? Are you concentrating on many aspects of the situation or focused on only one?"*, the female participants reported significantly lower values than their male counterparts (Table 2). Lower values in this case meaning that the female participants thought their concentration was not as divided nor did they feel like they had to concentrate hard, and that they did not feel like they had to concentrate on many different aspects during the experiment (Table 2).

Table 1 Pilot participant SART scores and task performance

#	Age	Gender	SART score	Answers	Events
1	27	M	11	7/10	14
2	29	M	16	7/10	13
3	29	F	5	2/10	0
4	26	F	7	3/10	0
5	31	M	11	4/10	4
6	27	F	14	1/10	13

Table 2 Pilot significant differences between genders

Attribute	Mean for men	Mean for women	*p*
SART Q5	5.33	2.33	0.033
SART Q6	5.67	2.00	0.005
Number of answers	6	2	0.026

p value threshold = 0.05

Table 3 Pilot SART score correlation with task performance metrics

Relationship	Pearson correlation	*p*
SART score and number of answers	0.421	0.406
SART score and logged events	0.852	0.031
SART score and total tasks (answers + events)	0.822	0.045

p value threshold = 0.05

When evaluating task performance, one category stands out as significant. The number of questions answered was significantly ($p = 0.026$) higher for the male group (Table 2). And while the differences in SART scores and total tasks completed (answers + events) were not found to be significant ($p = 0.279$ and $p = 0.153$), they still indicate that there could be significant differences to be found if the identical experiment was carried out with a bigger sample.

The correlation between the SART score and task performance (for all participants) showed significant results in two relationships. The correlation between SART score + logged events (Pearson correlation = 0.852, $p = 0.031$) and SART score + total tasks (answers + events) (Pearson correlation = 0.822, $p = 0.045$) revealed a significant, yet weak, correlation between the results (Table 3). These results indicate that the reported total SART score and task performance do in fact have a direct, albeit weak, correlation between them.

Additionally, when looking at the results, it can be seen that one participant in both gender groups performed differently to the other two. In the male group, one participant answered fewer questions and recorded fewer events than the other two, but still did so in equal amounts (4 answers and 4 events). In the female group, one concentrated singularly on the secondary event logging task and seemed to forget the main task of answering the questions. These findings are interesting because they do seem to indicate that in our experimental setup females only concentrated on a single task while males concentrated on both more equally.

5.2 Study

Like in the pilot, the task performance and SART scores were analyzed with one-way ANOVA. The new addition for the study, the NASA TLX, was also analyzed using one-way ANOVA. The majority of the analysis and comparisons were done between the male and female groups since this was the area that revealed significant statistical differences in the pilot. Due to the small sample size in this study, the non-parametric Kruskal-Wallis H test was used to further analyze the SART and NASA TLX results. Pearson correlation was used to analyze any correlation.

Task Performance This time there were four tasks for the participants to complete as opposed to the two in the pilot. In the pilot, the participants were told that only the main task was mandatory but this time all were mandatory but with varying levels of importance. Looking at the task performance (Table 4), we can observe that the participants actually spent time more equally between the tasks and did not just concentrate on one of them. One of the goals for the new system design and layout had been to make the system easier to use and multitask and it seems to have worked. The task performance saw some variation between men and women and also inside the gender groups, but the differences were not significant. The alarm that was implemented to indicate threat level 1 events worked really well and only one participant missed one of the alarms. The missed alarm was the first alarm of the session so it is likely they were still nervous or busy going over the AR functions. Despite there being many tasks to complete, one participant was still able to complete all of them in time. And even though this group of participants had 3 people who had tried the pilot as well, they did not fare any better in regards to the task performance than the 3 participants who had no prior experience with the system or even AR in general.

SART Scores Like in the pilot, there were no significant differences in the participants' overall SART scores. Comparing the same components where we saw significant differences in the pilot (Table 5), we can observe that this time the differences were not statistically significant. However, the trend is still the same, with men reporting a higher level of concentration and division of attention than women, and still scoring higher in overall task performance. Unlike in the task performance, there were some differences in the overall SART scores between the

Table 4 Participant task performance and SART score

#	Age	Gender	Task 1	Task 2	Task 3	Task 4	Total	SART
1	27	M	6/10	12/12	3/3	0/2	21/27	21
2	29	M	10/10	12/12	3/3	2/2	27/27	13
3	31	F	9/10	12/12	1/3	0/2	22/27	7
4	27	F	5/10	12/12	3/3	0/2	20/27	19
5	29	M	3/10	12/12	3/3	1/2	19/27	11
6	28	F	2/10	11/12	1/3	2/2	16/27	9

Table 5 SART differences between genders

Attribute	Mean for men	Mean for women	*p*
SART Q5	5.67	4.00	0.189
SART Q6	4.67	3.67	0.349
Tasks completed	22.33	19.33	0.745

p value threshold = 0.05

participants that partook in both experiments and the ones that only did the study. The people who had participated in the pilot experiment as well scored significantly higher ($p = 0.031$) in the overall SART score. Their mean score was 17.67 as opposed to 9.00 for the participants who only did the study experiment.

NASA TLX Statistical analysis of the task load index scores did not reveal any statistically significant differences between the genders. Though due to the low number of participants this is not a completely unexpected outcome. There were some interesting and visible differences in some categories but the *p*-value threshold was not reached in any of these cases. However, it could be useful to highlight some of these differences. For example, in the weight values for Performance and Frustration there was one participant in one or both groups that had a completely different value than the two others. In Performance, the values for men were: 3, 2, 5 and for women: 5, 5, 2. So one man and two women ranked Performance to the highest maximum weight and the others ranked it a much lower 2 or 3. For Frustration, the men had weights of 0, 0, 1 and women 4, 4, 0. There we can see a large variation in the women's weight ranking with two women ranking Performance as a really important factor and one reporting it basically did not matter at all. In many of the categories, the situation was exactly that there was one outlier in one or both groups that had enough of an impact so the *p*-value ended up above the threshold.

Pearson Correlation Running a correlation analysis revealed some interesting relationships in the data. In the NASA TLX results, there was a clear correlation between the Effort and the Performance categories (Table 6). All the parts (weight, raw rating and adjusted rating) of both categories correlated in a statistically significant manner (weight correlation 0.841 with $p = 0.036$, raw rating correlation 0.927 with $p = 0.008$ and adjusted rating correlation 0.895 with $p = 0.016$). These two categories seem to have the reverse values in all the categories. So for example, when Effort was given a high raw rating, the Performance received a low raw rating and vice versa. To clarify, the Performance category does not measure the actual performance of the user but rather how the user felt they succeeded in accomplishing the goals set by the researcher or by the participant themselves.

Unsurprisingly, the Temporal Demand category had a strong correlation with the Total Workload. The experiment had a time limit that was most likely slightly intimidating to the participants. The higher they evaluated the importance and scale of the Temporal Demand the higher their Total Workload ended up being. A similar kind of correlation was also observed between the adjusted rating for Frustration and Total Workload (though this fell slightly short of being significant p=0.053),

Table 6 Pearson correlation

Relationship	Pearson correlation	*p*
Performance weight/Effort weight	0.841	0.036
Performance raw rating/Effort raw rating	0.927	0.008
Performance adjusted rating/Effort adjusted rating	0.895	0.016
Temporal demand weight/Total tasks	0.913	0.011
Temporal demand adjusted rating/Total Tasks	0.914	0.011
Frustration weight/Total workload	0.872	0.023
Frustration adjusted rating/Total workload	0.805	0.053
Mental demand raw rating/SART Q2	0.873	0.023
Mental demand raw rating/SART Q7	0.814	0.048
Performance raw rating/SART Q5	0.952	0.003
Effort raw rating/SART Q5	0.894	0.016
Effort adjusted rating/SART Q6	0.846	0.034
Frustration weight/SART Q5	0.900	0.014
Frustration adjusted rating/SART Q5	0.903	0.014
Total tasks/SART Q7	0.850	0.032

p value threshold = 0.05

indicating that the less frustrated the participant was the lower their Total Workload was. There was no difference in Frustration between the users who had participated in the pilot and those who had not so the users that had more experience with the AR system, did not seem to have an unfair advantage in this case.

Many parts in the NASA TLX query also had a clear correlation with specific SART questions. For example Performance raw rating, Effort raw rating, and Frustration weight/adjusted rating correlated significantly with SART Q5. In this case the more Effort the participant reported, the higher the Q5 was. Reversely the higher Performance and Frustration (impact and rating) the user reported the lower the SART Q5 was. This means that more Effort equaled a higher concentration. But also the higher the reported Frustration and the worry about one's Performance was, the lower the reported concentration was.

The raw rating for Mental Demand also correlated with a few of the SART questions, significantly with Q2 and Q7. SART Q2 is about how complex the situation was and the results show a correlation of higher reported Metal Demand translating into higher reported situation complexity.

Kruskal-Wallis H Test Due to the low number of participants, the NASA TLX and SART results were also analyzed using the Kruskal-Wallis H test. There were no statistically significant findings when comparing the participants by gender. However, when doing the comparison by their experience with the systems (pilot + study vs. only study), a few statistically significant details were found (Table 7). The 3 participants with more experience reported a significantly higher level of Effort (raw rating) and significantly lower Performance "anxiety" (adjusted rating) than their less experienced counterparts. In regards to SART Q5 and overall SART score,

Table 7 Kruskal-Wallis significance between the group that took part in pilot + study vs. the group that only participated in the study

	SART Q5	SART	Effort raw rating	Performance adj. rating
K-W H	4.500	3.857	3.857	3.857
df	1	1	1	1
Asymp. sig.	0.034	0.05	0.05	0.05

p value threshold = 0.05

the more experienced ones reported a significantly higher value for Q5 and scored a significantly higher overall SA score. All these findings indicate that experience did have some impact in our test cases.

6 Discussion

In this user study, we aimed at observing the task performance using several metrics in a small-scale pilot and a follow-up experiment. We observed subtle gender differences that were significant using One-Way ANOVA but not with Kruskal-Wallis H Test.

Gender Differences in Task Performance In the pilot, we observed significant differences between males and females in their task performance. Especially in the main task of answering questions, males did better than females. For the follow-up study, the experiment structure was changed drastically, one major change being that now there were more tasks and all of them were mandatory but with varying levels of importance. So, in the study, the task performance became more equal between the gender groups and thus, the differences were not statistically significant. The three people who had prior experience of having been part of the pilot experiment as well, did not fare any better in task performance than the 3 new participants. So, it seems the new experiment model does not require previous AR experience for the user to be efficient at least in their tasks.

6.1 *Effect of Gender and Experience on SA*

In situational awareness overall scores, there were no significant differences between the genders in the pilot nor the study. The pilot though, revealed that men and women had significantly different answers to SART questions 5 and 6 and this was investigated again in the study. This time the differences were there but they were not statistically significant (SART Q5 $p = 0.0189$, SART Q6 $p = 0.349$). The pilot also showed correlation between overall SART score and some task elements but in the study there was no correlation found. Interestingly, albeit there was no difference in overall SART scores between genders, there was a statistically significant difference between the participants (3 people) who had done the pilot and the study and the users (3 people) who only did the study. The participants with more experience reported significantly higher SART scores than the ones who only participated in the latter study ($p = 0.031$). In future experiments, this difference should be taken into account when recruiting participants. It would probably be best to only include people who have not tested any version of the application before to rule out the effect of experience altogether.

6.2 Equal Workloads with Outliers

The NASA TLX was a new addition to the study and though there were no statistically significant differences found, some of the results were still interesting. In quite a few of the reported results there was one outlier in one or both of the gender groups that reported a drastically different category component (weight or raw rating) than the other two. Going forward these workload results should definitely be investigated further with a larger sample size to see if these outliers are just that: outliers, and if we get a more equal distribution of results.

6.3 Workload and SA Correlations

When analyzing for correlations in the data we found that Effort and Performance categories in the NASA TLX results correlated heavily with each other. These two had a reverse correlation in weight, raw rating, and adjusted rating. So when Effort had a high rating or weight, Performance would have an opposite low value and vice versa. What this means is that when the participant felt their effort level was low, their worry about their level of performance was high and interestingly when their effort was high they felt less stressed about performing well. We also found that Performance, Effort, and Frustration had a significant correlation with SART Q5. More specifically raw ratings for Performance and Effort and weight/adjusted rating for Frustration. In this case, Performance and Frustration had a reverse correlation with Q5 while Effort had a positive correlation. This makes sense because the more effort you put in the more you need to concentrate. And at the same time more frustration and performance anxiety you feel the less mental capacity you have to actually concentrate on the task. So, according to these results, it seems like when people feel frustrated and worry more about achieving certain goals, it impacts their situational awareness in a negative way, and at the same time the more effort they put in, the more positively it affects SA. Mental Demand also showed significant positive correlation with SART, mainly Q2 and Q7. Q7 asks the user to evaluate the mental capacity they have to spare and Q2 is about how complex the user feels the situation is. To interpret these results, a more complex situation is more demanding mentally but interestingly, with higher Mental Demand the participants still reported having more mental capacity to spare in the situation.

6.4 General Effects of Experience

Just as in SART and NASA TLX analysis, the Kruskal-Wallis H test did not reveal any differences between genders. However, significant differences were found when comparing the participants with more experience with the AR system to those with less experience. As mentioned previously, the overall SART score was significantly higher for the participants with more experience and this test again confirmed it but it also found a significant difference in SART Q5. The more experienced participants reported significantly higher concentration levels for the experiment. This could be because they already knew that the experiment requires their full attention and maybe this was something the new users did not realize. The better overall SART score could also be because the experienced users simply knew how to mentally prepare for the experiment better. This assumption is further reinforced by the fact that the experienced users also reported a significantly higher level of Effort (raw rating) and that their adjusted Performance rating was significantly lower than for the less experienced group. To sum up, the more experienced group concentrated more intensely, they had better situational awareness in general, they put in more effort, and worried less about achieving the goals set for the experiment. In simple terms it looks like having done the pilot, their knowledge and confidence were higher and it showed in these results. So, perhaps in the next experiment, it might be a good idea to include only people that have not done either of these experiments so we can rule out the experience part when analyzing the results.

6.5 Limitations and Future Work

We had a small sample size in both experiments meaning we cannot generalize our results. Despite referring to the participants' gender in the above we do not propose there are conclusive gender differences based on our study. Our experiment should be repeated again with more improvements to the testing procedure, a more polished implementation, and also with a larger sample size to validate or debunk some of the more interesting findings and to gain statistically significant results; we also aim to rule out the impact of prior experience by recruiting from the public for follow-up studies.

We used One-Way ANOVA for analysis, with a small sample without being able to tell whether the sample follows normal distribution. A non-parametric test was not able to verify the results, but actually showed a statistical difference between experienced and non-experienced participants. This highlights that our results are inconclusive, however, they suggest interesting paths for future research.

Because of the potential bias of XR technologies, both hardware and software, this is a valid target for future research. Both hardware and software influence FOV and the relative situation of user interface (UI) elements. It is possible that due to these differences, solutions for dynamically situating UI items for optimal attention are needed. While almost all current commercial VR and AR HMDs have a calibration setup that also includes the IPD, the UI design and situating elements on the display are still subject to design and the whim of the developers.

6.6 Ethics Statement

This research has been conducted following the ethical requirements established by the Finnish National Board on Research Integrity (TENK) [34] and the guidelines provided by the Ethics Committee of Human Sciences of the University of Oulu [35]. All the collected material has been handled with respect to privacy and anonymity in accordance with Finnish laws. Informed consent was acquired from each study participant before the experimental trial.

7 Conclusions

We conducted two experiments, a pilot and a study, to measure SA in a multitasking environment, where AR was used for system monitoring while other tasks were being done using more traditional means such as a desktop computer and pen and paper. The main takeaway from the pilot results is that there were some significant, albeit subtle, differences in how women and men performed in this experiment. The task performance in the pilot revealed that women focused on one activity while males dispersed their attention across tasks. Interestingly, women still reported needing less focus and having their attention split less than males. These results could not be verified with the second study and there were no significant differences in how the men and women performed in the tasks. Our study has several limitations due to the small sample size. These results contradict current literature on attention and task performance which is why it is quite possible that the root cause is the hardware, software, and design of the available smart glasses.

In the pilot, the relationships between SART scores and two task performance metrics also showed significant correlation indicating that better situational awareness does translate to increased task performance. While there are other studies suggesting gender differences in aspects of task performance in multitasking situations, there are also assuring results showing that there are no differences [21, 36], which is an additional reason for remaining wary of the results we are seeing in our study. Again in the succeeding study, the differences between the genders were not significant.

The most interesting findings in the study were the different correlations between NASA TLX and SART scores and also how some categories in the NASA TLX correlated with each other. The differences between the more experienced and less experienced users was also an interesting discovery.

Acknowledgments This work has been partially funded by the European Commission grants NESTOR (101021851), PRINCE (815362) and IDUNN (101021911); Business Finland project Reboot Finland IoT Factory 33/31/2018; and Academy of Finland 6Genesis Flagship (318927).

References

1. Ferreira, A., Cruz-Correia, R.: Covid-19 and cybersecurity: finally, an opportunity to disrupt? JMIRx Med **2**(2), e21069 (2021)
2. Pranggono, B., Arabo, A.: Covid-19 pandemic cybersecurity issues. Internet Technol. Lett. **4**(2), e247 (2021)
3. Setianto, F., Tsani, E., Sadiq, F., Domalis, G., Tsakalidis, D., Kostakos, P.: Gpt-2c: a parser for honeypot logs using large pre-trained language models. In: Proceedings of the 2021 IEEE/ACM International Conference on Advances in Social Networks Analysis and Mining, pp. 649–653 (2021)
4. Garzón, J., et al.: Systematic review and meta-analysis of ar in educational settings. Virtual Real. **23**(4), 447–459 (2019)
5. Aschenbrenner, D., et al.: Artab-using vr and ar methods for an improved situation awareness for telemaintenance. IFAC-PapersOnLine **49**(30), 204–209 (2016)
6. Irizarry, J., et al.: Infospot: a mobile ar method for accessing building information through a situation awareness approach. Autom. Constr. **33**, 11–23 (2013). https://doi.org/10.1016/j.autcon.2012.09.002
7. Korkiakoski, M., Sadiq, F., Setianto, F., Latif, U.K., Alavesa, P., Kostakos, P.: Using smart glasses for monitoring cyber threat intelligence feeds. In: Proceedings of the 2021 IEEE/ACM International Conference on Advances in Social Networks Analysis and Mining, pp. 630–634 (2021)
8. Kumar, A., et al.: Trends in existing and emerging cyber threat intelligence platforms (2019). https://doi.org/10.35940/ijiteeL31881081219
9. Sauerwein, C., et al.: Threat intelligence sharing platforms: an exploratory study of software vendors and research perspectives (2017)
10. Wagner, C., et al.: Misp: the design and implementation of a collaborative threat intelligence sharing platform. In: Proceedings of the 2016 ACM on Workshop on Information Sharing and Collaborative Security, pp. 49–56. WISCS '16, Association for Computing Machinery, New York, NY, USA (2016). https://doi.org/10.1145/2994539.2994542
11. MISP: Covid-19 misp information sharing community (2019). https://www.misp-project.org/covid-19-misp/
12. Vinot, R., Dulaunoy, A., Luxembourg: CIRCL - Computer Incident Response Center Luxembourg, Van Impe, K. PyMISP (2022). https://github.com/MISP/PyMISP
13. Fonnet, A., Prie, Y.: Survey of immersive analytics. IEEE Trans. Vis. Comput. Graph. **27**(3), 1–22 (2019)
14. Slater, M.: A note on presence terminology. Presence Connect **3**(3), 1–5 (2003)
15. Kabil, A., Duval, T., Cuppens, N., Le Comte, G., Halgand, Y., Ponchel, C.: 3d cybercop: a collaborative platform for cybersecurity data analysis and training. In: International Conference on Cooperative Design, Visualization and Engineering, pp. 176–183. Springer (2018)
16. Blackler, A., Hurtienne, J.: Towards a unified view of intuitive interaction: definitions, models and tools across the world. MMI-Interaktiv **13**(2007), 36–54 (2007)

17. Ens, B., Bach, B., Cordeil, M., Engelke, U., Serrano, M., Willett, W., Prouzeau, A., Anthes, C., Büschel, W., Dunne, C., et al.: Grand challenges in immersive analytics. In: Proceedings of the 2021 CHI Conference on Human Factors in Computing Systems, pp. 1–17 (2021)
18. Kress, B.C., Chatterjee, I.: Waveguide combiners for mixed reality headsets: a nanophotonics design perspective. Nanophotonics **10**(1), 41–74 (2021)
19. knob2001: Oculus quest 2: AnÃlisis (Oct 2020), https://www.realovirtual.com//articulos/5646/oculus-quest-2-analisis
20. Dodgson, N.A.: Variation and extrema of human interpupillary distance. In: Stereoscopic Displays and Virtual Reality Systems XI, vol. 5291, pp. 36–46. International Society for Optics and Photonics, Bellingham (2004)
21. Grassini, S., Laumann, K.: Are modern head-mounted displays sexist? a systematic review on gender differences in hmd-mediated virtual reality. Front. Psychol. **11**, 1604 (2020)
22. Kostakos, P., Alavesa, P., Korkiakoski, M., Marques, M.M., Lobo, V., Duarte, F.: Wired to exit: exploring the effects of wayfinding affordances in underground facilities using virtual reality. Simul. Games **52**(2), 107–131 (2021)
23. Lindemann, P., et al.: Supporting driver sa for autonomous urban driving with an ar windshield display. In: 2018 IEEE International Symposium on Mixed and Augmented Reality Adjunct (ISMAR-Adjunct), pp. 358–363. IEEE (2018)
24. Hong, T.C. et al.: Assessing the sa of operators using maritime augmented reality system (Mars). In: Proceedings of the Human Factors and Ergonomics Society Annual Meeting, vol. 59, pp. 1722–1726. SAGE Publications, Los Angeles (2015)
25. Pascale, M.T., et al.: The impact of head-worn displays on strategic alarm management and sa. Hum. Factors **61**(4), 537–563 (2019)
26. Coleman, J., Thirtyacre, D.: Remote pilot sa with ar glasses: an observational field study. Int. J. Aviat. Aeronaut. Aerosp. **8**(1), 3 (2021)
27. Gans, E., et al.: Ar technology for day/night situational awareness for the dismounted soldier. In: Display Technologies and Applications for Defense, Security, and Avionics IX; and Head- and Helmet-Mounted Displays XX, vol. 9470, p. 947004. International Society for Optics and Photonics (2015)
28. Rowen, A., et al.: Impacts of wearable augmented reality displays on operator performance, situation awareness, and communication in safety-critical systems. Appl. Ergon. **80**, 17–27 (2019)
29. Aschenbrenner, D., et al.: An exploration study for ar and vr enhancing situation awareness for plant teleanalysis. In: International Design Engineering Technical Conferences and Computers and Information in Engineering Conference, vol. 58110, p. V001T02A065. American Society of Mechanical Engineers (2017)
30. Göbel, M.: SimpleJSON (2021). https://wiki.unity3d.com/index.php/SimpleJSON
31. Taylor, R.M.: Situational awareness rating technique (sart): the development of a tool for aircrew systems design. In: Situational Awareness, pp. 111–128. Routledge, London (2017)
32. Salmon et al., P.M.: Measuring sa in complex systems: Comparison of measures study. International Journal of Industrial Ergonomics **39**(3), 490–500 (2009)
33. NASA TLX links: (2022). https://humansystems.arc.nasa.gov/groups/tlx/tlxpaperpencil.php
34. TENK: Finnish National Board on Research Integrity (TENK) (2022). https://tenk.fi/en
35. University of Oulu: Ethics committee of human sciences (2022). https://www.oulu.fi/en/university/faculties-and-units/eudaimonia-institute/ethics-committee-human-sciences
36. Hirsch, P., et al.: Putting a stereotype to the test: the case of gender differences in multitasking costs in task-switching and dual-task situations. PloS One **14**(8), e0220150 (2019)

Effects of Global and Local Network Structure on Number of Driver Nodes in Complex Networks

Abida Sadaf, Luke Mathieson, and Katarzyna Musial

Abstract Control of complex networks is one of the most challenging open problems within network science. One view says that we can only claim to fully understand a network, if we have the ability to influence or control it and predict the results of the employed control mechanisms. Investigating and understanding global network structures like network density, centrality measures or shortest paths and local structures like communities is an important space in many domains and disciplines, including the spread of news on social networks. To be able to develop more efficient control mechanisms, we need to understand the relationship between different global and local network structures and the number of driver nodes needed to control a given structure. This will allow understanding of which networks might be easier to control and the resources needed to control them. In this paper, we present an experimental study that investigates how number of driver nodes identified in communities of the network can relate to the densities of those communities, and also we discuss the difference between number of driver nodes in overall network versus number of driver nodes in communities. For this purpose networks are generated using various network models (random (R), small-world (SW), scale-free (SF)). Moreover, we also consider twenty-two real social networks to examine, if the results from generated networks can be confirmed in the real world scenarios.

Keywords Social networks · Network models · Network structure measures · Number of driver nodes · Communities · Complex networks · Control in complex networks

---

A. Sadaf (✉) · L. Mathieson · K. Musial
Complex Adaptive Systems Lab, School of Computer Science, University of Technology Sydney, Sydney, NSW, Australia
e-mail: Abida.Sadaf@student.uts.edu.au

S. T. Özyer, B. Kaya (eds.), *Cyber Security and Social Media Applications*, Lecture Notes in Social Networks, https://doi.org/10.1007/978-3-031-33065-0_4

1 Introduction

A complex system can be represented by a complex network, such as represented in [1–5]. To be able to control a complex network is a critical endeavour and is still relatively an unexplored research direction [6]. The notion is that if we control, every single node in the network could be fully controllable. This approach however is not very cost-effective, because traversing through every node requires abundant resources and also is not feasible in many scenarios [7, 8]. Hence, control in a complex network is defined by detecting the minimum number of driver nodes needed to control the whole system. To identify these driver nodes in the network, an algorithm based upon maximum matching principle [9] has been explored before [10].

Synthetic models of networks and real world networks have been an active part of the complex network research for many years. Many interesting studies have been carried out in the directions of network modelling and evolution, control over a network [11–13], information diffusion, link prediction [14] and community detection [15–19].

Due to the extreme use of online social networking sites, social networks are studied at great length in network science. In social networks, there exist complex interactions among numerous unique nodes. Finding communities from the social network is a difficult task because of its structure that is complex, dynamically changing and the communities themselves can be overlapping [20]. In case of community detection, the methods to detect communities have been applied to identify terrorists organisations [17], recommending products for users [18], anomaly detection [15, 19], finding potential friends in social media [16] and analysing social opinions [21] to name a few.

From review of the previous work, it is clear that we do not know if and how the network structure correlates with the number of driver nodes [22]. As driver nodes play a key role in achieving control of a complex network, identifying them and studying their correlation with network structure measures can bring valuable insights, such as what network structures are easier to control, and how can we alter the structure in our favour to achieve the maximum control over the network. Our previous research work [22] determines the relationship between some of the global network structure measures and number of driver nodes. A systematic study presented in [22] builds an understanding of how global network profiles of synthetic (random, small-world, scale-free) and real social networks influence the number of driver nodes needed for control. In our previous work we focused on global structural measures such as network density and how it can play an important role in determining that how big or small the identified driver nodes set will be. Our results show that as density increases in networks like random, small world and scale free, the number of driver nodes tends to decrease. That is how we were able to determine the network structures that were easier to control.

Our current work focuses on both global and local structural measures and their relationship with number of driver nodes. We propose that communities (as being local structure measure) are found to be one of the most important features of networks, and detecting them enables us to analyse and explore further underlying structural features of the synthetic as well as social networks. The idea is to detect communities and driver nodes within the communities to see how the number of communities influences the number of driver nodes. Based upon the review of the previous research work, we have formulated the following research challenge. Identifying a set of driver nodes in complex networks has always been very vital in control of a complex network. The full control of social networks is very hard to achieve due to their varying structures, dynamics and complex human behaviour that is hard to control completely. However, we can start by detecting the potential relationships between number of driver nodes and underlying structural measures (global and local). Our main research questions for this work are stated as follows:

- **Is there a relationship between community density and number of driver nodes found in the communities of the networks?**
- **Is there a difference in number of driver nodes found in the overall network as compared to when detected in communities?**

This document contains the following sections: Sect. 2 describes related work and the main research challenge that is the focus of this study. Sections 3 and 4 describe (1) the research methodology and experimental design in detail and (2) include results and analysis of the experiments performed respectively. Finally, the conclusions drawn from the experiments, future work and limitations are discussed in Sects. 5 and 6 respectively.

2 Background

In social networks, identification of a minimum set of driver nodes is a potential research problem. This paper contains the research work that focuses on understanding how global and local network structures relate to the minimum number of driver nodes [13, 23, 23–30]. In complex networks, small or large communities within the network are organically present. There are many definitions of communities in the networks, but in general they may be defined as a group of nodes which are densely connected with the other nodes in the group and sparsely connected to the nodes outside that group [31]. From our previous research work, as in [22], we identified the network structures that were easier to control from the perspective of global measures. We found out that networks with higher density require smaller number of driver nodes to control the underlying network. For that, we calculated structural measures for three types of generated networks and their profiles. Generated networks that were used in experimentation were random, small world and scale free networks. We also extended the experiment to include real social networks, for that purpose we analysed 22 social networks [22]. More details about these networks can be found in Sect. 3.

2.1 Community Detection

There are many community detection algorithms in use, for example GN [31], FN [31], CNM [31], LPA [32], EM [33], SCAN [34], Louvain [35], LFM [36], Infomap [37, 38] and NMF [39] to name a few.

We start by utilising the widely used and tested algorithm, The Girvan-Newman algorithm (GN Algorithm). A brief description of the algorithm is given below, and an example of identified communities by using GN Algorithm from Zachare Karate Club is presented in Fig. 1. The Girvan–Newman algorithm detects communities by progressively removing edges from the original graph. The algorithm removes the edge with the highest betweenness centrality, at each step. As the graph breaks down into pieces, a community structure is exposed, and the result can be represented as a dendrogram. Below is the step-by-step process of the GN Algorithm.

1. Create a network of N nodes and its edges.
2. Calculate the betweenness of all existed edges in the network.
3. Remove all the edges with the highest betweenness.
4. Recalculate the betweenness of all the edges that got affected by the removal of edges.
5. Repeat steps 3 and 4 until no edges remain [31].

2.2 Driver Nodes

Previously, some models/methods have been proposed for the identification of driver nodes that can potentially control the complex networks, for example; interbank networks [23], protein interaction networks [24], biological networks [23, 25–27], and

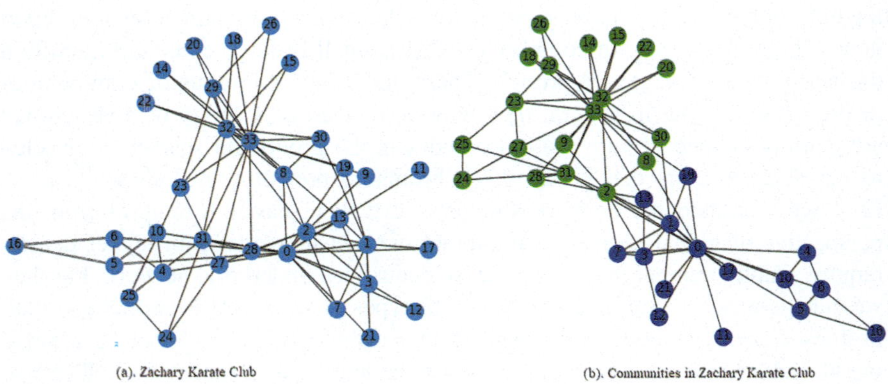

Fig. 1 Representation of Zachary karate club network and detected communities. (**a**) Zachary karate club. (**b**) Communities in Zachary karate club

real networks [13, 28–30]. Effects of local and global network structural measures including degree distribution have been explored in [40]. Many methods/algorithms have been proposed to identify a set of number of driver nodes from a network. These methods can potentially be called as control methods. The control methods are based upon these algorithms Maximum Matching Algorithm [9], Hopcroft-Karp Algorithm [9], Hungarian Algorithm [41] and Minimum Dominating Set [10] have been previously used to identify a set of driver nodes from an underlying network. Despite these algorithms, a complexity of choosing a smaller set of driver nodes still exists. It means, given this number, the largest possible subset of the network can be controlled. If we have to restrict to this smaller set, we should have a ranking of driver nodes that allow us to pick those that have the largest impact on controlling the network. Existing measures for such a ranking, for example control capacity [42], and control range [7], are not best suited because they only focus on one aspect of driver nodes, either their probability to become a driver or the size of the subnetwork they control. Control contribution combines both of these two aspects [29]. In [29], researchers have claimed that control contribution, will always be able to efficiently and effectively control the network, however, this requires further evaluation, and extensive studies, to determine exactly, which method/measures will be efficiently and effectively control a complex network. Below is a description of Minimum Dominating Set (MDS) method. We have used this method to identify driver nodes from within networks and communities, both. MDS method is a state-of-the-art method to identify driver nodes from any network. A *dominating set* of a graph G is a set S of nodes, where every node in G is either an element of S or adjacent to an element of S. A *minimum dominating set (MDS)* is a dominating set of the smallest cardinality. MDS approach allows us to identify a minimum dominating set as the set of driver nodes such that the network is made structurally controllable. Because the nodes in the minimum dominating set as the set of driver nodes as each node is either a driver node, or adjacent to a driver node from which it can receive a control signal [10], see Fig. 2.

We can see from Fig. 2 that the network can be fully controlled by selecting an MDS i.e., a set of driver nodes because each dominated node has its own control signal. The maximum matching approach needs three driver nodes a, b, and d,

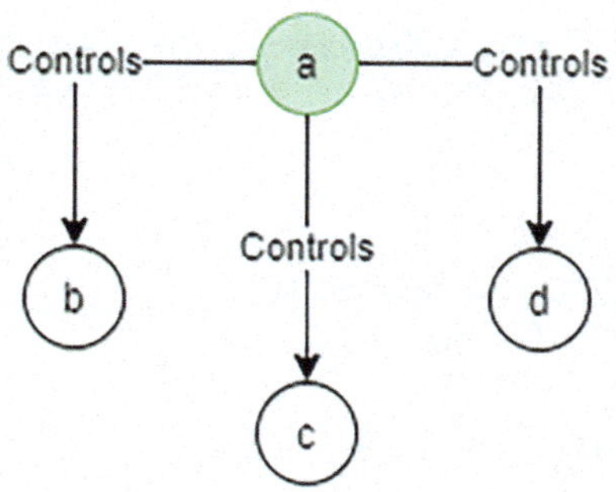

Fig. 2 Minimum dominating set (MDS) model, adapted from [22]

assuming a matching link from a to c. However, the MDS only requires one node, i.e., a. Where a can assume control of b, c and d [10].

3 Research Methodology

Figure 3 explains the methodology of the research work. In our previous work, the work revolves around the correlation between number of driver nodes and global network structure measures such as network density, as published in [22]. While in the extended work, we focus on global as well as local structure measures. So, we identify communities in the network by using GN community detection algorithm and then identifying driver nodes within those communities to see which network structures become easier to control in terms of less number of driver nodes. Following is the experiment design of this research work. Following are a series of new experiments that were conducted to achieve the research questions mentioned in Sect. 2.

1. Detecting communities by using GN algorithm for all the generated and social networks.
2. A generalised framework has been developed which is suitable to detect communities from all networks. Framework, identify the nodes in the communities as different sets. Communities can also be presented as dendrograms.

We mean to analyse the following results from the above experiments. Firstly, correlation between community densities and number of driver nodes in those communities. Secondly, number of communities and number of driver nodes within those communities is identified and see if the driver nodes set is bigger or smaller

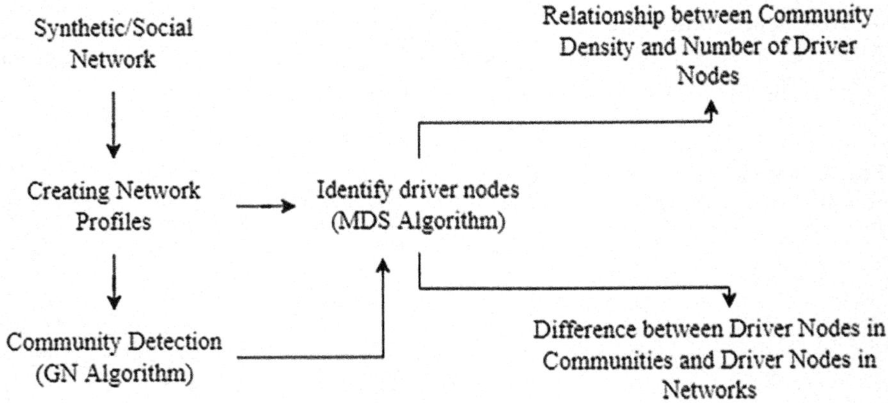

Fig. 3 Methodology

than the set achieved through simply by using Minimum Dominating Set method to identify number of driver nodes in the networks [10]. For the research questions identified in Sect. 2 and experiments proposed above, the following experimental setup has been adopted to conduct the experiments.

1. Generate random, small world and scale free networks. Ten networks of each node size of 100, 200, 300, 400 and 500 are generated with varying densities i.e., small, medium and large. Densities of these networks ranges from minimum density of 0.16 to the medium density of 0.55 to the highest density of 1.
2. The experiments mean to extend to include real social networks in the final analysis. The datasets have been downloaded from Stanford.[1]
3. Calculate structural measures i.e., network densities (D) in generated as well as real networks as presented in Tables 3 and 1.
4. Identification of number of driver nodes (NDN) in overall networks by using MDS method in both generated and social networks. NDN densities for generated networks are given in Table 1 and NDN for social networks are presented in Table 3.
5. Analysed the relationship between global structural measures, i.e., density with NDN as presented in [22].
6. Identify communities in synthetically generated networks and in social networks, the global structure measures of which are presented in Table 3. We utilised NetworkX library of Python programming language to generate networks. GN Algorithm has been used to detect communities in the network. A step-by-step process is given in Sect. 2. The algorithm and setup has been implemented in Python version 3.6. Also, the algorithm used a said threshold to stop, which is defined as the square root of the number of nodes in the network.
7. Identify driver nodes in communities in synthetic and social networks. We used Minimum Dominating Set Algorithm for identifying the driver nodes in communities. A description of the algorithm is given in Sect. 2.
8. Correlation between community densities and number of driver nodes is done by obtaining densities of the communities and identifying number of driver nodes in those communities by MDS method.
9. The difference (Diff.) between total number of driver nodes identified in overall networks (NDN) as compared to the number of driver nodes found in communities of those networks (NDNC) is also obtained by obtaining results partially from the previous study [22], and largely from the current one. The Diff. tells us, the significance of identifying driver nodes within communities, like following a divide and conquer approach.

[1] https://snap.stanford.edu/data/.

Table 1 Global network structure measures i.e., number of nodes, number of edges and network density ND with their corresponding number of driver nodes density i.e. $N_d D$ in random, small world and scale free networks

	Random			Small world			Scale-free		
Nodes	Edges	ND	$N_d D$	Edges	ND	$N_d D$	Edges	ND	$N_d D$
100	800	0.18	0.162	800	0.178	0.16	800	0.22	0.149
	1600	0.09	0.323	1600	0.095	0.32	1600	0.17	0.272
	2400	0.06	0.485	2400	0.068	0.48	2400	0.12	0.368
	3200	0.05	0.646	3200	0.045	0.65	3200	0.09	0.44
	4000	0.03	0.808	4000	0.033	0.81	4000	0.07	0.56
	4800	0.02	0.97	4800	0.019	0.97	4800	0.05	0.71
	4950	0.01	1	4950	0.01	1	4950	0.03	0.877
200	2400	0.133	0.12	2400	0.135	0.121	2400	0.145	0.113
	4800	0.076	0.24	4800	0.08	0.241	4800	0.14	0.212
	7200	0.054	0.36	7200	0.05	0.362	7200	0.125	0.367
	9600	0.04	0.48	9600	0.045	0.499	9600	0.095	0.463
	12,000	0.028	0.6	12,000	0.025	0.603	12,000	0.075	0.482
	14,400	0.023	0.724	14,400	0.02	0.724	14,400	0.06	0.567
	16,800	0.017	0.844	16,800	0.015	0.844	16,800	0.04	0.654
	19,200	0.012	0.965	19,200	0.01	0.965	19,200	0.025	0.787
	19,900	0.005	1	19,900	0.005	1	19,900	0.02	0.898
300	12,800	0.047	0.285	12,800	0.05	0.288	12,800	0.09	0.337
	19,200	0.031	0.428	19,200	0.03	0.428	19,200	0.083	0.366
	22,400	0.027	0.499	22,400	0.027	0.502	22,400	0.063	0.392
	25,600	0.023	0.571	25,600	0.03	0.569	25,600	0.053	0.441
	28,800	0.02	0.642	28,800	0.017	0.642	28,800	0.05	0.456
	32,000	0.013	0.713	32,000	0.013	0.742	32,000	0.047	0.428
	35,200	0.013	0.785	35,200	0.01	0.789	35,200	0.037	0.502
	38,400	0.012	0.856	38,400	0.01	0.856	38,400	0.027	0.569
	41,600	0.01	0.928	41,600	0.007	0.93	41,600	0.023	0.642
	44,850	0.003	1	44,850	0.003	1	44,850	0.017	0.742
400	40,000	0.02	0.501	40,000	0.023	0.501	40,000	0.045	0.301
	44,000	0.015	0.551	44,000	0.018	0.551	44,000	0.038	0.351
	48,000	0.015	0.602	48,000	0.015	0.602	48,000	0.033	0.401
	52,000	0.015	0.652	52,000	0.015	0.652	52,000	0.03	0.451
	60,000	0.013	0.752	60,000	0.013	0.752	60,000	0.028	0.501
	64,000	0.01	0.802	64,000	0.01	0.802	64,000	0.023	0.675
	68,000	0.01	0.852	68,000	0.008	0.852	68,000	0.01	0.802
	72,000	0.008	0.902	72,000	0.005	0.902	72,000	0.008	0.852
	76,000	0.005	0.952	76,000	0.005	0.952	76,000	0.005	0.902
	98,000	0.003	1	98,000	0.003	1	98,000	0.005	0.952

(continued)

Table 1 (continued)

Nodes	Random			Small world			Scale-free		
	Edges	ND	N_dD	Edges	ND	N_dD	Edges	ND	N_dD
500	72,000	0.018	0.577	72,000	0.018	0.569	72,000	0.024	0.404
	76,800	0.014	0.616	76,800	0.014	0.613	76,800	0.022	0.436
	81,600	0.012	0.654	81,600	0.012	0.653	81,600	0.016	0.5
	86,400	0.01	0.693	86,400	0.01	0.693	86,400	0.016	0.681
	91,200	0.01	0.731	91,200	0.01	0.729	91,200	0.016	0.721
	96,000	0.01	0.77	96,000	0.01	0.754	96,000	0.018	0.771
	100,800	0.008	0.808	100,800	0.008	0.806	100,800	0.014	0.816
	105,200	0.006	0.843	105,200	0.006	0.842	105,200	0.012	0.842
	110,000	0.006	0.882	110,000	0.006	0.882	110,000	0.01	0.882
	124,750	0.002	1	124,750	0.002	1	124,750	0.01	0.898

4 Results and Analysis

4.1 *Results from Synthetic Networks*

In this section, we analyse the results obtained from the conducted experiments. Below are the comparisons, that have been carried out to answer the research question. Some network structure measures related to network structure measures of random, small world and scale free networks are given in Table 1 along with number of driver nodes density (N_dD). The table has been adapted from our previous work presented in [22]. Following are the important results and their analysis.

Community Density and Number of Driver Nodes in Communities in Random, Small World and Scale Free Networks Figure 4, we correlated local structure measures such as community density with number of driver nodes within those communities. It is evident that as community densities start to approach 1, so are driver nodes. We can see that as community densities are higher, the number of drive nodes are low and vice versa. This result is answers our first research question and

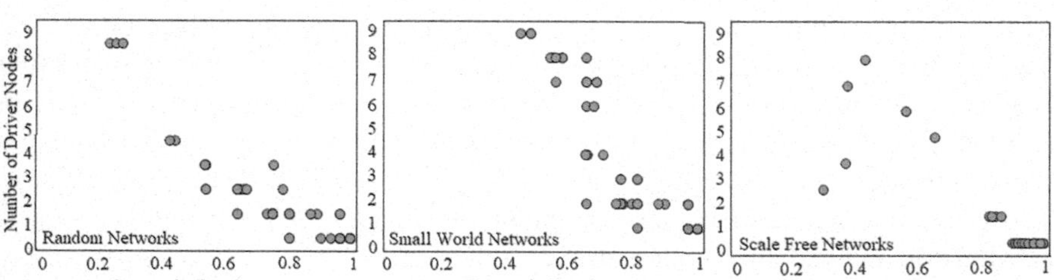

Fig. 4 Number of driver nodes in the communities of random, small world and scale free networks verses their community densities

also strengthens the results from previous paper, where we structurally correlated the global measures i.e., network density with number of driver nodes [22].

Difference Between Number of Driver Nodes in Networks (NDN) and Number of Driver Nodes in Communities (NDNC) in Random, Small World and Scale Free Networks Firstly, from Table 1 it is clear that we are able to minimise the number of driver nodes (NDN) in the overall network, as we are able to increase the number of edges in all three generated networks. By increasing the number of edges, we automatically increase the density of the network. We can also see that there is not much difference between the density of number of driver nodes $N_d D$ in the networks, as all networks were able to minimise the number of driver nodes to 1 with increased edges and density. More details of these results are provided in [22]. From Fig. 5a, b and c, we can see the difference between NDN (number of driver nodes) and NDNC (number of driver nodes detected in communities) in random, small world and scale free networks. We can see a big difference between the plots of scale free networks in comparison to random and small world networks. From the figure, the conclusions from [22] again strengthens that, as the density tend to increase i.e., number of edges increase in the same node size network, number of driver nodes tend to decrease within the network as well as within the communities of those networks. For example, in a scale free network of (nodes = 400 and edges = 79,000), only 1 driver node is required within the communities of that network. It is because network itself is really very dense, and communities are naturally denser than network itself by definition. Secondly, we know that, structurally random, small world and scale free networks are quite different from each other. We applied the correlation analysis on our generated network which indicates that, random, small world and scale free networks are quite different from each other. They behaved differently when identifying driver nodes from communities. This can be seen from Fig. 5. Thirdly, Table 2 a huge difference between the number of driver nodes within the whole network as compared to within the communities of those networks. The table shows a heatmap of the difference between NDN and NDNC (Diff.). We can see from the map that the most difference is found in scale free networks and least difference is found in small world networks, while in random networks the difference is kind of in the middle. We figure that, most real world networks have scale free properties, that is why difference is larger in the social networks as can be seen from Table 3. This inference strengthens our observations from the experiments. Lastly, Fig. 6 shows the number of communities in all generated networks. We can see that there is more variation in number of communities in scale free networks as compared to random and small world networks despite the same network sizes. As scale free networks are analysed to be closer in structure to real networks, they can have more communities in networks as compared to their small world and scale free counterparts.

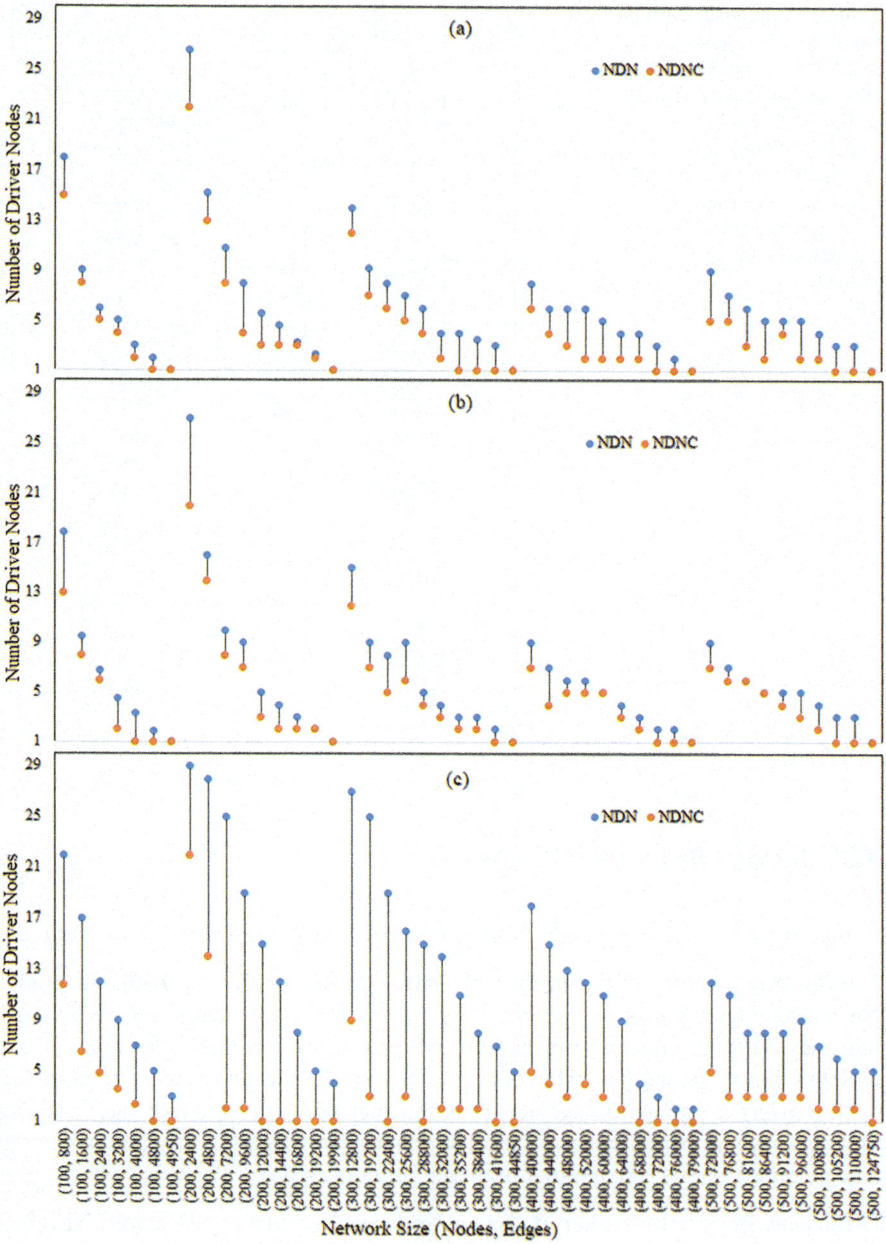

Fig. 5 Difference between number of driver nodes in overall network verses number of driver nodes in communities found in (**a**) Random networks, (**b**) Small world networks and (**c**) Scale free networks

Table 2 Difference between Number of Driver Nodes (NDN) in the Whole Network and Number of Driver Nodes within the Communities (NDNC) of the Networks, i.e., random (R), small world (SW) and scale free (SF) (Diff.). Nodes and Edges of the networks are also presented

Nodes	Edges	Diff.						Edges	Nodes
		R	SW	SF	R	SW	SF		
	800	3	5	9	5	7	9	2400	
	1600	1	2	9	2	2	10	4800	
	2400	1	1	6	3	2	17	7200	
100	3200	1	3	7	4	2	12	9600	
	4000	1	2	6	3	2	12	12000	200
	4800	1	1	4	2	2	10	14400	
	4950	0	0	2	0	1	6	16800	
	12800	2	3	15	0	0	3	19200	
	19200	2	2	18	0	0	3	19900	
	22400	2	3	14	2	2	11	40000	
	25600	2	3	10	2	3	11	44000	
300	28800	2	1	11	3	1	8	48000	
	32000	2	1	11	4	1	7	52000	
	35200	3	1	9	3	0	6	60000	400
	38400	3	1	6	2	1	6	64000	
	41600	2	1	6	2	1	2	68000	
	44850	0	0	4	2	1	2	72000	
	72000	4	2	5	1	1	1	76000	
	76800	2	1	5	0	0	1	798000	
500	81600	3	0	2	2	2	5	100800	
	86400	3	0	3	2	2	5	105200	500
	91200	1	1	4	2	2	4	110000	
	96000	3	2	2	0	0	4	124750	

4.2 Results from Social Networks

In this section, results and analysis from social networks are discussed in detail.

Community Density and Number of Driver Nodes in Communities in Social Networks From Fig. 7 it is evident that communities, by definition, have high densities. This confirms our results from previous study that the denser the network (or community, as we have shown here), the smaller the number of driver nodes [22]. This is a strong reassurance showing that network structure has a strong influence on number of driver nodes. The same figures clearly show that, when the density approaches 1, the number of driver nodes decrease.

Difference Between Number of Driver Nodes in Networks (NDN) and Number of Driver Nodes in Communities (NDNC) in Social Networks We calculated the difference between total number of driver nodes in the whole network (NDN) versus number of driver nodes in the communities of the networks (NDNC). The difference of both (Diff.) indicates that in all the social networks, number of driver nodes decrease when they are identified within communities. It strongly indicates that, the divide and conquer approach works for the networks. Also, it is easier

Table 3 Social networks and their global and local structure measures such as nodes (N), edges (E), density (D), number of driver nodes (NDN), number of communities (C), number of driver nodes in communities (NDNC) and difference between number of driver nodes in networks and number of driver nodes in communities (Diff.)

Networks	N	E	D	C	NDN	NDNC	Diff.
FB [43]	4039	88,234	0.01	180	499	270	229
Z [44]	34	78	0.14	2	13	9	4
Twitter [43]	23,371	32,832	0.00012	350	939	489	450
Diggs [45]	192,4000	3,298,475	0.000002	156,432	398,004	199,037	198,967
Youtube [46]	1,134,891	2,987,625	0.000004	54,983	136,520	68,285	68,235
Ego [43]	23,629	39,195	0.00014	75	132	96	36
LC [47]	4658	33,116	0.003	517	1178	620	558
LF [48]	874	1309	0.0034	97	347	209	138
PF [49]	1858	12,534	0.0073	206	745	398	347
MFb [50]	22,470	171,002	0.00067	2643	11,955	6011	5944
DHR [51]	54,574	498,202	0.0003	6420	15,678	7877	7801
DRO [51]	41,774	125,826	0.0001	4914	22,680	11,372	11,308
DHU [51]	47,539	222,887	0.0002	5592	29,479	14,755	14,724
MG [50]	37,700	289,003	0.0004	4435	15,507	7775	7732
L [52]	7624	27,806	0.0009	759	3518	1795	1723
FbAR [51]	50,516	819,306	0.0006	5943	28,670	14,372	14,298
FbA [51]	13,867	86,858	0.0009	1383	6827	3449	3378
FbG [51]	7058	89,455	0.0036	784	4245	2160	2085
FbN [51]	27,918	206,259	0.0005	3284	15,558	7813	7745
FbP [51]	5909	41,729	0.0024	562	2995	1530	1465
FbPF [51]	11,566	67,114	0.001	1051	5510	2792	2718
FbT [51]	3893	17,262	0.0023	387	1966	1011	955

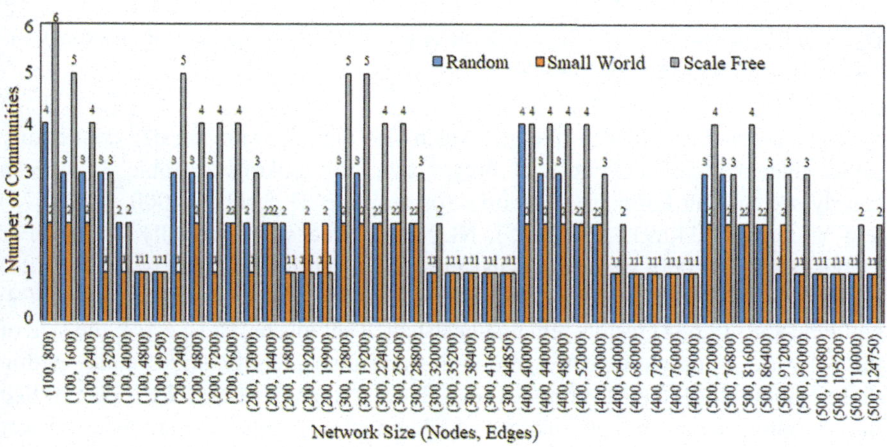

Fig. 6 Number of communities in random, small world and scale free networks

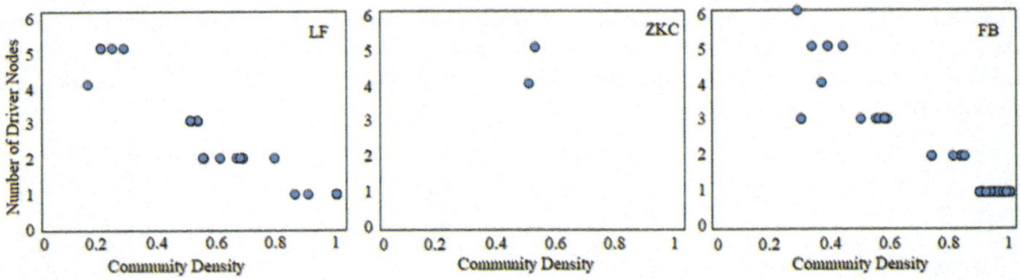

Fig. 7 Number of driver nodes in the communities of LF, Z and FB networks verses their community densities

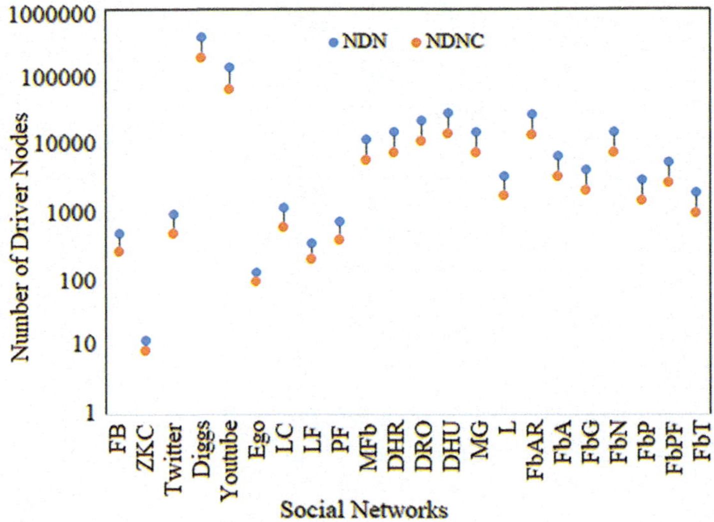

Fig. 8 Difference between total number of driver nodes (NDN) and number of driver nodes in communities (NDNC) in social networks

to apply the process of identifying driver nodes within a smaller size community rather than a huge network with bigger size. By looking at Table 3, we can clearly see that, in large size networks for example, Diggs(1,924,000, 3,298,475) and Youtube(1,134,891, 2,987,625), NDN set reduces substantially in size. For Diggs, NDN(481,000) becomes NDNC(198,967), and for Youtube, NDN(283,722) becomes NDNC(68,235). Even for small networks like ZKC(34, 78), LF(874, 1309) and PF(1858, 12,534) results remain consistent. That means, irrespective the size of the network, when we detect communities and then identify driver nodes within those communities, the driver nodes set reduces to a great extent. We observed that in communities, densities are relatively higher naturally, hence the number of driver nodes in communities is less than the number of driver nodes in their corresponding overall networks.

We can see an overall picture for all the social networks from Fig. 8, where the plot shows the difference of huge amount between NDN and NDNC values within the social networks. Since we do not have overlapping communities due to the nature of the algorithm, we have at least one driver node within each community.

5 Discussion and Conclusion

One of the key findings in [22], was that the global network structural measures (i.e., community densities) do indeed correlate with the number of driver nodes found in those communities. When the values of the investigated structural measures increase or decrease, this directly triggers the increase or decrease in the number of driver nodes. From our previous work, we found out that the denser the network, the smaller the number of driver nodes, meaning and those network structures are easier to control [22]. Through this study, we answer the research questions stated in Sect. 2. Communities themselves have pretty large densities as compared to the overall network. So, connecting from the previous results, it seems only plausible that, within those communities, we will find a minimum number of driver nodes with more potential to control the community and by controlling those communities, ultimately the overall network.

Our main contributions in this research work are given below:

1. First key contribution is the study of finding relationships between local structure measures of the network and the number of driver nodes. This has not been explored before. From this research, we contribute that local structure measures such as community densities potentially correlated with the decrease or increase in number of driver nodes. It is easier to control the communities with higher densities because these communities potentially have less number of driver nodes and in some cases equal to 1, when density also approaches 1.
2. By detecting driver nodes within communities, we potentially decrease the total number of driver nodes as were identified in the whole network. Hence, it is recommended to break the network down in communities to conquer the problem of identifying a minimum set of driver nodes. We can clearly see from Figs. 5 and 8, that there is a difference in number of driver nodes when identified within communities as compared to when identified in overall (i.e. random, small world, scale free and real social) networks. This potential result can help the researchers with the problem of identifying an optimal set of driver nodes that can control the networks.
3. MDS method to detect driver nodes is a very expensive process in very large networks, specially real social networks [10]. By dividing the networks in communities, we potentially can reduce the size of the overall network in smaller subnetworks, which makes the process of identifying driver nodes comparatively less time-consuming. By presenting this idea, we open another dimension to minimise the driver nodes set to a certain point, which can still remain effective in controlling the overall network. Our method can positively guide the research in control of complex networks, where we might be able to get an efficient set

of driver nodes to control the overall network. We stress to make this point that, global as well as local structural measures of the networks can play an important role to figure out an efficient way to determine the potential driver nodes set that can control a social network.

In conclusion, network structure measures, global and local, do indeed play an important role in determining the minimum number of driver nodes. In future, many more analyses can be done with different other kinds of networks with varying new structural measures to see the potential correlations. This can open the avenues for making the process of identifying a minimum set of driver nodes an approachable and feasible idea.

6 Limitations and Future Work

Although a comparative analysis in [53], tells us that GN algorithm might not be suitable for large networks because of its high computation time. But, it has never been actually tested with large scale networks. However, the computation simplicity of the algorithm gives it an advantage over the more complex algorithms, which makes it a potential candidate to employ it in this research work. Despite the high computation time that it takes to identify the communities in the network, the idea of allowing to identify no overlapping communities make it easier to identify unique driver nodes within the communities of the network later on. High computation costs of determining communities by applying GN algorithm and identifying driver nodes by MDS method pose a limitation for this study. As we need a huge amount of resources to use these methods for very large scale networks. For future work, new and more cost-effective methods of community detection can be employed to further decrease the overall cost of the process.

Acknowledgments This work was supported by Australian Research Council, Dynamics and Control of Complex Social Networks, under Grant DP190101087.

References

1. Pastor-Satorras, R., Vespignani, A.: Epidemic spreading in scale-free networks. Phys. Rev. Lett. **86**(14), 3200 (2001)
2. Gao, Z.K., Jin, N.D.: A directed weighted complex network for characterizing chaotic dynamics from time series. Nonlinear Anal. Real World Appl. **13**(2), 947–952 (2012)
3. Gao, Z.K., Fang, P.C., Ding, M.S., Jin, N.D.: Multivariate weighted complex network analysis for characterizing nonlinear dynamic behavior in two-phase flow. Exp. Thermal Fluid Sci. **60**, 157–164 (2015)
4. Gao, Z.K., Yang, Y.X., Fang, P.C., Jin, N.D., Xia, C.Y., Hu, L.D.: Multi-frequency complex network from time series for uncovering oil-water flow structure. Sci. Rep. **5**(1), 1–7 (2015)

5. Luo, J., Qi, Y.: Identification of essential proteins based on a new combination of local interaction density and protein complexes. PloS One **10**(6), e0131418 (2015)
6. Liu, B., Chu, T., Wang, L., Xie, G.: Controllability of a leader–follower dynamic network with switching topology. IEEE Trans. Autom. Control **53**(4), 1009–1013 (2008)
7. Wang, B., Gao, L., Gao, Y.: Control range: a controllability-based index for node significance in directed networks. J. Stat. Mech. Theory Exp. **2012**(04), P04011 (2012)
8. Chen, Y.Z., Wang, L., Wang, W., Lai, Y.C.: The paradox of controlling complex networks: control inputs versus energy requirement. Preprint (2015). arXiv:1509.03196
9. Hopcroft, J.E., Karp, R.M.: An n^5/2 algorithm for maximum matchings in bipartite graphs. SIAM J. Comput. **2**(4), 225–231 (1973)
10. Nacher, J.C., Akutsu, T.: Dominating scale-free networks with variable scaling exponent: heterogeneous networks are not difficult to control. New J. Phys. **14**(7), 073005 (2012)
11. Liu, Y.Y., Barabási, A.L.: Control principles of complex systems. Rev. Mod. Phys. **88**(3), 035006 (2016)
12. Ding, J., Lu, Y.Z.: Control backbone: an index for quantifying a node s importance for the network controllability. Neurocomputing **153**, 309–318 (2015)
13. Burbano-L, D.A., Russo, G., di Bernardo, M.: Pinning controllability of complex stochastic networks. IFAC-PapersOnLine **50**(1), 8327–8332 (2017)
14. Martínez, V., Berzal, F., Cubero, J.C.: A survey of link prediction in complex networks. ACM Comput. Surv. (CSUR) **49**(4), 1–33 (2016)
15. Wang, J., Paschalidis, I.C.: Botnet detection based on anomaly and community detection. IEEE Trans. Control Netw. Syst. **4**(2), 392–404 (2016)
16. Zhu, J., Wang, B., Wu, B., Zhang, W.: Emotional community detection in social network. IEICE Trans. Inf. Syst. **100**(10), 2515–2525 (2017)
17. Saidi, F., Trabelsi, Z., Ghazela, H.B.: A novel approach for terrorist sub-communities detection based on constrained evidential clustering. In: 2018 12th International Conference on Research Challenges in Information Science (RCIS), pp. 1–8. IEEE (2018)
18. Li, C., Zhang, Y.: A personalized recommendation algorithm based on large-scale real micro-blog data. Neural Comput. Appl. **32**(15), 11245–11252 (2020)
19. Keyvanpour, M.R., Shirzad, M.B., Ghaderi, M.: Ad-c: a new node anomaly detection based on community detection in social networks. Int. J. Electron. Bus. **15**(3), 199–222 (2020)
20. Sathiyakumari, K., Vijaya, M.: Community detection based on girvan newman algorithm and link analysis of social media. In: Annual Convention of the Computer Society of India, pp. 223–234. Springer, Berlin (2016)
21. Wang, D., Li, J., Xu, K., Wu, Y.: Sentiment community detection: exploring sentiments and relationships in social networks. Electron. Commerce Res. 17(1), 103–132 (2017)
22. Sadaf, A., Mathieson, L., Musial, K.: An insight into network structure measures and number of driver nodes. In: Proceedings of the 2021 IEEE/ACM International Conference on Advances in Social Networks Analysis and Mining, pp. 471–478 (2021)
23. Delpini, D., Battiston, S., Riccaboni, M., Gabbi, G., Pammolli, F., Caldarelli, G.: Evolution of controllability in interbank networks. Sci. Rep. **3**, 1626 (2013)
24. Wuchty, S.: Controllability in protein interaction networks. Proc. Natl. Acad. Sci. **111**(19), 7156–7160 (2014)
25. Liu, Y.Y., Slotine, J.J., Barabási, A.L.: Controllability of complex networks. Nature **473**(7346), 167 (2011)
26. Guo, W.F., Zhang, S.W., Wei, Z.G., Zeng, T., Liu, F., Zhang, J., Wu, F.X., Chen, L.: Constrained target controllability of complex networks. J. Stat. Mech. Theory Exp. **2017**(6), 063402 (2017)
27. Guo, W.F., Zhang, S.W., Zeng, T., Li, Y., Gao, J., Chen, L.: A novel structure-based control method for analyzing nonlinear dynamics in biological networks, p. 503565 (2018). bioRxiv
28. Zhang, J.X., Chen, D.B., Dong, Q., Zhao, Z.D.: Identifying a set of influential spreaders in complex networks. Sci. Rep. **6**, 27823 (2016)
29. Zhang, Y., Garas, A., Schweitzer, F.: Control contribution identifies top driver nodes in complex networks. Preprint (2019). arXiv:1906.04663

30. Hou, W., Ruan, P., Ching, W.K., Akutsu, T.: On the number of driver nodes for controlling a boolean network when the targets are restricted to attractors. J. Theor. Biol. **463**, 1–11 (2019)
31. Girvan, M., Newman, M.E.: Community structure in social and biological networks. Proc. Natl. Acad. Sci. **99**(12), 7821–7826 (2002)
32. Raghavan, U.N., Albert, R., Kumara, S.: Near linear time algorithm to detect community structures in large-scale networks. Phys. Rev. E **76**(3), 036106 (2007)
33. Newman, M.E., Leicht, E.A.: Mixture models and exploratory analysis in networks. Proc. Natl. Acad. Sci. **104**(23), 9564–9569 (2007)
34. Xu, X., Yuruk, N., Feng, Z., Schweiger, T.A.: Scan: a structural clustering algorithm for networks. In: Proceedings of the 13th ACM SIGKDD International Conference on Knowledge Discovery and Data Mining, pp. 824–833 (2007)
35. Blondel, V.D., Guillaume, J.L., Lambiotte, R., Lefebvre, E.: Fast unfolding of communities in large networks. J. Stat. Mech. Theory Exp. **2008**(10), P10008 (2008)
36. Lancichinetti, A., Fortunato, S., Kertész, J.: Detecting the overlapping and hierarchical community structure in complex networks. New J. Phys. **11**(3), 033015 (2009)
37. Rosvall, M., Bergstrom, C.T.: Maps of random walks on complex networks reveal community structure. Proc. Natl. Acad. Sci. **105**(4), 1118–1123 (2008)
38. Yiapanis, P., Rosas-Ham, D., Brown, G., Luján, M.: Optimizing software runtime systems for speculative parallelization. ACM Trans. Archit. Code Optim. **9**(4), 1–27 (2013)
39. Zhang, S., Wang, R.S., Zhang, X.S.: Uncovering fuzzy community structure in complex networks. Phys. Rev. E **76**(4), 046103 (2007)
40. Sorrentino, F.: Effects of the network structural properties on its controllability. Chaos Interdisciplinary J. Nonlinear Sci. **17**(3), 033101 (2007)
41. Kuhn, H.W.: The hungarian method for the assignment problem. Naval Res. Logist. Q. **2**(1–2), 83–97 (1955)
42. Jia, T., Barabási, A.L.: Control capacity and a random sampling method in exploring controllability of complex networks. Sci. Rep. **3**, 2354 (2013)
43. McAuley, J.J., Leskovec, J.: Learning to discover social circles in ego networks. In: NIPS, vol. 2012, pp. 548–56. Citeseer, Princeton (2012)
44. Zachary, W.W.: An information flow model for conflict and fission in small groups. J. Anthropol. Res. **33**(4), 452–473 (1977)
45. Hogg, T., Lerman, K.: Social dynamics of digg. EPJ Data Sci. **1**(1), 1–26 (2012)
46. Yang, J., Leskovec, J.: Defining and evaluating network communities based on ground-truth. Knowl. Inf. Syst. **42**(1), 181–213 (2015)
47. Kunegis, J.: Konect: the koblenz network collection. In: Proceedings of the 22nd International Conference on World Wide Web, pp. 1343–1350 (2013)
48. Guo, G., Zhang, J., Thalmann, D., Yorke-Smith, N.: Etaf: an extended trust antecedents framework for trust prediction. In: 2014 IEEE/ACM International Conference on Advances in Social Networks Analysis and Mining (ASONAM 2014), pp. 540–547. IEEE (2014)
49. Rossi, R., Ahmed, N.: The network data repository with interactive graph analytics and visualization. In: Proceedings of the AAAI Conference on Artificial Intelligence, vol. 29 (2015)
50. Rozemberczki, B., Allen, C., Sarkar, R.: Multi-scale attributed node embedding. J. Complex Netw. **9**(2), cnab014 (2021)
51. Rozemberczki, B., Davies, R., Sarkar, R., Sutton, C.: Gemsec: Graph embedding with self clustering. In: Proceedings of the 2019 IEEE/ACM International Conference on Advances in Social Networks Analysis and Mining, pp. 65–72 (2019)
52. Rozemberczki, B., Sarkar, R.: Characteristic functions on graphs: birds of a feather, from statistical descriptors to parametric models. In: Proceedings of the 29th ACM International Conference on Information & Knowledge Management, pp. 1325–1334 (2020)
53. Yang, Z., Algesheimer, R., Tessone, C.J.: A comparative analysis of community detection algorithms on artificial networks. Sci. Rep. **6**(1), 1–18 (2016)

A Lightweight Global Taxonomy Induction System for E-Commerce Concept Labels

Mayank Kejriwal and Ke Shen

Abstract Given a domain-specific set of concept labels as input, *global taxonomy construction* (GTC) is defined as the problem of automatically inducing a taxonomy over the concept labels. Despite its importance in domains such as e-commerce and healthcare, and recent algorithmic research as a result, practical tools for inducing and interactively visualizing taxonomies over domain-specific concept sets do not currently exist. To be truly useful, such a tool must permit a reasonable solution in a relatively unsupervised setting, and be applicable to general subsets of concept labels. In this paper, we present InVInT (Interactive Visualization and Induction of Taxonomies), an unsupervised and lightweight global taxonomy induction system for arbitrary concept-labels. We demonstrate the utility of InVInt on e-commerce concept labels taken from challenging real-world datasets. The system is end-to-end, only taking a simple text file as input and yielding a tree-like taxonomy as output that can then be rendered on a browser, and that a non-technical practitioner can interact with. Important components of the system can also be customized by a technically experienced user.

Keywords Global taxonomy induction · E-commerce · Product taxonomy · Representation learning · Visualization · D3 · Transfer learning

---

The authors "Mayank Kejriwal" and "Ke Shen" contributed equally to this work.
This paper is an extended version of [1], which was an 4-page demonstration paper presented and published in the 2021 IEEE/ACM ASONAM conference (held virtually). In this article, we significantly expand upon the workings of the underlying Taxonomy Induction over Concept Labels (TICL) algorithm, present more analysis and additional visualizations, an evaluation, and a new section on discussion and limitations.

M. Kejriwal (✉) · K. Shen
Information Sciences Institute, University of Southern California, Los Angeles, CA, USA
e-mail: kejriwal@isi.edu; keshen@isi.edu

S. T. Özyer, B. Kaya (eds.), *Cyber Security and Social Media Applications*,
Lecture Notes in Social Networks, https://doi.org/10.1007/978-3-031-33065-0_5

1 Introduction

Structured representations of concepts, including graphs and taxonomies ('trees'), have always played an important role in various contexts and domains, especially in applications that offer front-facing users many choices, such as online shopping and navigation [2, 3], and price comparisons [4]. This is especially evident in a digital world, when users have short attention span and face a deluge of information and advertisements from different websites. Beyond Web-oriented contexts, such as e-commerce, domain-specific taxonomies play an important role in various fields and applications e.g., scientific workflows [5], as well as healthcare [6].

While in some cases, a taxonomy is 'standard' in the field and used by professionals and practitioners in that field, on e-commerce websites (including the websites for Target, Walmart and even Google Shopping [7]), different taxonomies are in place, despite the marketplaces selling many of the same products. On online platforms where customers make purchase decisions, a well-designed taxonomy can have a positive impact on several Web-oriented applications. Not only can it serve as a structured representation of the 'categories of products' (such as 'baby diapers' or 'black beans') that the platform is selling, but the taxonomy is also useful for website navigation and search [8]. A well-designed taxonomy, according to which the website's pages are linked and organized, can make for a more seamless online experience. However, the taxonomy cannot be static, since media and e-commerce companies typically have to organize many labels that tend to evolve as new products, brands, product variations and topics (for media companies) are introduced into online marketplaces and forums.

Given these many use-cases, inducing and interactively visualizing a taxonomy in a relatively unsupervised fashion over a set of concept-labels is an important and practical problem, especially for Web enterprises. In this chapter, we present an end-to-end tool called InVInT (Interactive Visualization and Induction of Taxonomies) that makes this task more frictionless, potentially allowing non-technical domain experts, product managers and marketers to spend their time more productively while still getting the pragmatic benefits of lightweight, automatic tools. The input to InVInT is simple: a text file that contains a set of concept-labels or 'phrases', while the output is an induced taxonomy that can be visualized and interacted with on a browser.

As subsequently discussed, we have validated the system on real-world, widely used e-commerce concept-label sets for which ground-truth taxonomies are available (and hence were used for illustrating the efficacy of the system in a rigorous and quantifiable way) but we will also allow the user to input their own files into the system. Next, we use the state-of-the-art Taxonomy Induction over Concept Labels (TICL) algorithm that uses minimally supervised transfer learning techniques from the machine learning literature to induce a taxonomy over the concept-labels set without any supervision. The system does this by relying on pre-trained representation learning algorithm and background resources from the NLP and Web sciences communities, respectively. Finally, the induced taxonomy is rendered visually on a browser using open-source tools. The user can manipulate this visualization and explore the induced taxonomy. To the best of our knowledge, this is the first such openly published tool that is not only able to induce a taxonomy (without any example links) in near-real time but also renders it in a visually appealing format for subject matter experts to navigate.

The rest of this chapter is structured as follows. In Sect. 2, we discuss related work and background on relevant topics, such as representation learning and taxonomy induction. Next, in Sect. 3, we formalize the problem of global taxonomy construction and briefly describe our previously implemented approach to the problem, called TICL (Taxonomy Induction over Concept Labels) that has already been validated against competitive baselines in an earlier article [9]. We use real-world examples to demonstrate both its utility and its limitations, and also provide some quantitative results showcasing its effectiveness in the e-commerce domain in particular. We then describe the workflow of the full InVInT system in Sect. 4, discussing its effectiveness using a preliminary experiment that resembles an A/B-test. We follow this with a discussion in Sect. 5 on the relevance of the presented results, followed by limitations and future work. Section 6 concludes the chapter.

2 Related Work

The advent of deep neural networks in the previous decade has led to significant advances in representation learning or 'embeddings' [10–13], which have come to become prominent in several applied AI and machine learning domains, including Natural Language Processing (NLP). Over the last five years, transformer-based neural networks such as BERT and GPT-3 obtained state-of-the-art performance on several NLP tasks [11, 14, 15]. Despite being expensive to train, with many requiring ever-growing corpora and billions (and even trillions) of parameters, than previous word embedding algorithms (such as GloVE [12] and FastText [13]) needed, pre-trained versions of these algorithms have since been released in the larger community and can be 'fine-tuned' for specific application tasks and domains. Examples of such tasks include text summarization, Named Entity Recognition, question answering, and many others [16–20].

Beyond words, sentences and linguistic representations, graphs have also been embedded by similar approaches (the most recent ones of which have also been transformer-based). Examples of network embeddings include DeepWalk and node2vec [21, 22], while so-called 'knowledge graph embeddings' include RDF2vec [23], popular approaches such as TransE and HolE [24], and domain-specific neural embeddings [25]. One issue with these approaches is that they assume that the set of vertices (or what is referred to as the concept-set in this chapter) is known before test-time; furthermore, there is an expectation of being given a set of 'positive' or example links between some of the vertices, which is then typically used to train a model and perform graph 'completion' by inferring other edges, and possibly eliminating noisy edges. Neither assumption holds here.

While our proposed system also depends on embeddings, especially in the natural language sense, it is able to work on generic e-commerce datasets with hundreds, and even thousands, of concept-labels without requiring any example edges. Furthermore, we provide an interactive visualization of the induced taxonomy. In the NLP literature, interactive visualizations (especially generated in an on-demand fashion on reasonably challenging datasets) continue to be rare, and the focus is usually on offline (or 'trained') validation on pre-decided benchmark datasets.

Although taxonomy (and more generally, *ontology*) induction has witnessed some research in domain-*generic* contexts, one limiting assumption is that the underlying data on which the taxonomy is induced tends to be context-rich, and often involves a set of generic nouns (e.g., WordNet extensions) [26], if not an accompanying text corpus, such as in the *OntoLearn Reloaded* system [27]. Although such induction approaches are highly beneficial for applications where text corpora are sufficient in that the text itself can provide the necessary context for the induction process, they cannot be applied if a text corpus is not available to begin with.

Algorithmic approaches best related to our underlying induction algorithm (TICL) include MaxTransGraph [28], MaxTransForest [29] and TAXI [30]. Similar to the algorithm used in InVInT, most of these approaches also assume a graph theoretic view of the global taxonomy construction problem. TAXI [30], for example, uses a series of patterns to extract the candidate hypernym relations between concepts from domain-specific corpora, and prunes the hypernym candidates per term. Another notable example is MaxTransGraph [28], which defined a directed graph structure over predicates representing entailment relations, and then modeled the global transitivity constraint within an Integer Linear Programming framework to infer the optimal set of directed edges. Similarly, MaxTransForest aims to scalably learn "transitive graphs" containing tens of thousands of vertices, with the vertices representing predicates (and edges representating "entailment rules"). However, none of these algorithms is specifically optimized for e-commerce data and none can generate interactive visualizations.

The system we describe is easily executed as a Docker container and renders a visualization of the taxonomy with only a small lag that can be interacted with on a browser. It was demonstrated recently in the 2021 ASONAM conference's demonstration track [31]. To the best of our knowledge, this is the first openly published architecture to do so. In other work, we also showed that our underlying algorithm (TICL) outperformed these (more generic) baselines on e-commerce data. Additional evaluation results in this chapter show that, in a blind A/B-test like evaluation comparing TICL outputs to a baseline, while controlling for the visualization component of InVinT, annotators still prefer TICL to the baseline. A more quantitative evaluation comparing TICL trees to 'ideal' gold-standard trees provided by annotators also shows a much smaller margin of error compared to the baseline.

3 Global Taxonomy Construction (GTC): Formalism and Approach

Global taxonomy construction (GTC) may be stated as the problem of inducing a taxonomy T given a concept set C, potentially along with some background or 'training' resource B. The induced taxonomy $T = \{(c_i, c_j) \mid c_i, c_j \in C \times C, c_i \neq c_j\}$ can be defined in more general ways but in web-oriented domains, especially e-commerce, it is usually defined as a tree-like structure to guide product categorization and website navigation. Another example is healthcare, with its well-organized taxonomies of terms that both medical providers and insurance companies can rely on (e.g., for deciding which services to bill and reimburse a patient for). A concrete example is the International Classification of Diseases 11th Revision (ICD-11) [6], which serves as a "global standard for diagnostic health information" according to the World Health Organization. Specifically, a concept in the induced taxonomy can have only one parent, thereby meeting the definition of a tree. Formally, if $(c_i, c_j) \in T$ and $(c_k, c_j) \in T$, then $c_i = c_k$, c_i is called as the parent of c_j. To orient the tree, we assume that the concepts in the top level have the artificial 'ROOT' node as their parent.

Due to combinatorial explosion,[1] the building process of a complete taxonomy itself is a difficult problem, and the internal error-cascading[2] problem is also challenging to address. In the literature, researchers [31] have instead focused on pragmatic approximations, such as an information retrieval (IR)-based methodology that only partially addresses the issue. They aimed to determine the local neighborhood (including the super-type and the sub-type) of each concept $c \in C$ in the unknown taxonomy T in a piecewise fashion, rather than induce the full taxonomy over C. Essentially, they described the problem they were solving as local taxonomy construction (LTC). Although the LTC system is helpful (since the global taxonomy could be recovered if the local 'fragments' in the LTC were combined together in a systematic way), an end-to-end GTC system that efficiently yields a reasonably high-quality taxonomy in an unsupervised fashion has still not been achieved yet, to the best of our knowledge. Evaluating the quality of rival taxonomies is also a challenging problem, since intuitively, users do not view such trees as collections of edges. Although one can compare to a 'gold-standard' taxonomy, multiple taxonomies of equally good quality might exist, further compounding the problem. These are some reasons why interactive visualization forms an important component of the InVInT architecture. Previous architectures, such as MaxTransForest, MaxTransGraph and TAXI, did not include interactive visualization but it is theoretically possible for their implementations

[1] The number of possible trees in the number of concepts is super-polynomial.

[2] In essence, this means that errors early in the construction process could have a multiplying effect on the quality of the entire tree.

to be integrated into InVInT due to a modular separation between the taxonomy induction and interactive visualization steps, as subsequently described.

3.1 Taxonomy Induction Over Concept Labels (TICL)

To address some of the challenges discussed above, we recently developed the *Taxonomy Induction over Concept Labels* (TICL) algorithm specifically for global taxonomy construction (GTC) [9]. We briefly describe it here, with complete details and evaluations provided in our journal article [9]. The core approach of TICL is to combine a *retrofitted* word embedding model from the NLP community with a graph-theoretic *spanning tree* algorithm to achieve good performance without seeing any training links [32, 33]. To accomplish this task robustly for arbitrary e-commerce concept-label sets, the algorithm draws upon a combination of background resources, including pre-trained word embedding models [13] as well as other publicly available taxonomies on the Web, such as *Google Product Category* [7] and *PriceGrabber* [4], for transfer learning.

The input of TICL is composed of three parts: a target concept set that is used to induce the taxonomy, the pre-trained word embedding model that contains the word embeddings of target concepts, and an example taxonomy. The algorithm also needs a parameter k that represents the maximum degree of each node in the induced taxonomy. The first step of the algorithm is to retrofit the pre-trained embedding based on the example taxonomy. If two concepts in the example taxonomy were linked, for instance, the embeddings of these two concepts will be pushed closer in the retrofitted embedding space. After its conclusion, this step yields the retrofitted embeddings of concepts in the input concept set.

In the second step, using these embeddings, TICL first constructs a 'draft' taxonomy as an undirected, approximately k-regular graph[3] by finding the k closest concepts of every single node in the retrofitted embedding space. Note that each node in this graph represents a concept in the input concept set. The reason that the graph is only *approximately* k-regular graph is that, for some nodes, it may be the case that they are connected to nodes of which they themselves are not included as k nearest neighbors. For example, assume a node n_1 that is connected to its k nearest neighbors, and a node n_2, which is not connected to n_1. Suppose it is the case that n_1 is in the k-nearest neighbors set of n_2. In this case, we would end up adding an edge between n_1 and n_2, when we construct the k nearest neighbors of n_2 and link each of those neighbors to n_2 (if they are not already linked). In the example above, n_1 will now have at least $k + 1$ neighbors. In practice, however, almost all nodes have approximately k neighbors.

To create the tree-like taxonomy and avoid excessive linkages, TICL uses a classic *minimum spanning tree* [33] technique to find a 'best' tree in the approximately

[3] The k-regular graph is a graph that every node has exactly k neighbors.

k-regular graph. In the final step, TICL randomly selects a node and appends an artificial ROOT node to it to orient the taxonomy as a top-down hierarchy. We provide a good and bad example result of TICL respectively in Fig. 1, given two different example concept sets. When the concepts came from a similar or the same category, as shown in the second subplot in Fig. 1, TICL outputs a taxonomy that accurately captures the hierarchical relationships between concepts. In the induced taxonomy, customers can find dog supplies, cat supplies, bird supplies, fish supplies,

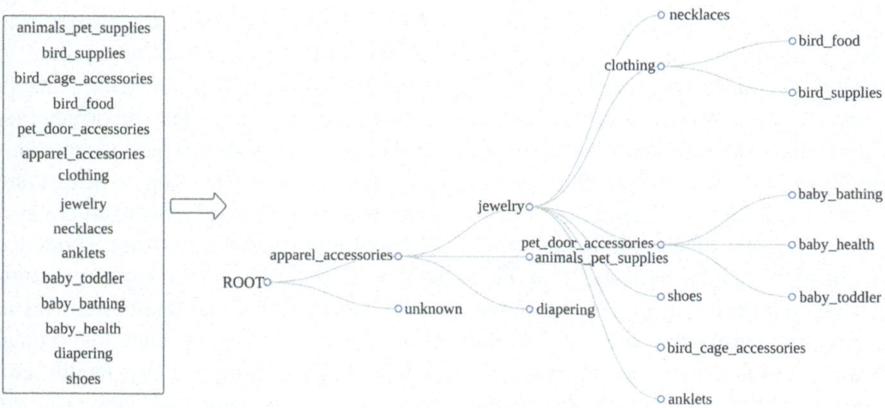

(a) Example where TICL produces a lower quality taxonomy

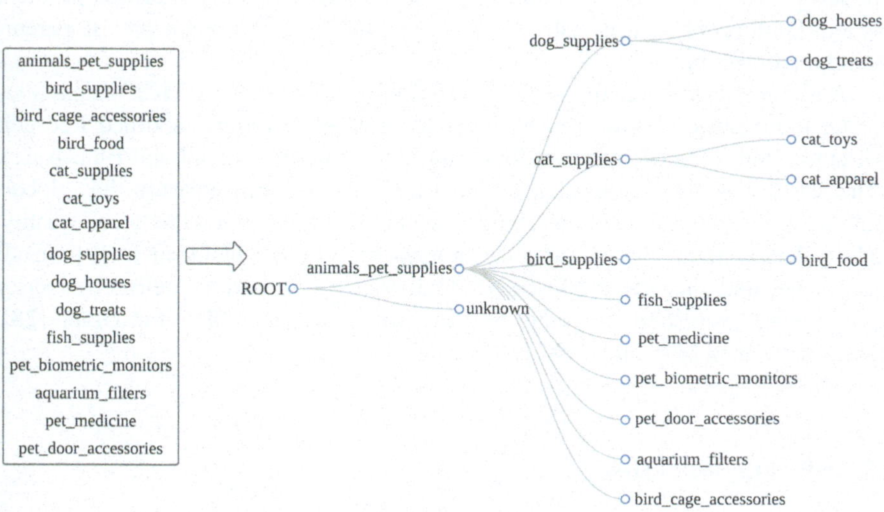

(b) Example where TICL produces a higher quality taxonomy

Fig. 1 The TICL results for two example-sets of 15 e-commerce terms derived from real-world datasets. (**a**) Example where TICL produces a lower quality taxonomy. (**b**) Example where TICL produces a higher quality taxonomy

pet medicine, pet biometric monitors, and pet door accessories in the animals and pet supplies category. TICL also correctly classifies the 'dog_houses' and the 'dog_treats' as dog supplies. Similarly, the 'bird_food' belongs to bird supplies and the 'cat_apparel' and the 'cat_toys' are cat supplies. However, the 'aquarium_filters' and the 'bird_cage_accessories' should be classified as fish supplies and bird supplies, respectively. TICL misclassifies them both as animals and pet supplies. It is still reasonable but imprecise.

A much worse case is shown in the first subplot in Fig. 1, when a few concepts are input from several different categories. Specifically, we input 15 e-commerce concepts across 3 categories: *animal and pet supplies, apparel accessories* and *baby / toddler*. Since TICL randomly chooses only one concept to connect to ROOT, it could not correctly link all three concepts to the ROOT. Only the concept 'apparel_accessories' was selected as the first-level category. The incorrect first level affects the assignment of concepts in the next few levels. The concept 'animal_pet_supplies' and 'diapering' were misclassified as 'apparel_accessories', and 'bird_food' and 'bird_supplies' were misclassified as 'clothing.' This example also illustrates one of the challenges of GTC (potentially cascading errors). Evidence from these results indicates that TICL pays more attention to the semantic similarities between concepts when inducing the taxonomy, and can sometimes fail to correctly identify the parent-child relationships between concepts. The most typical example of this phenomenon is when TICL puts 'baby_bathing', 'baby_health' and 'baby_toddler' in the same level, under the 'pet_door_accessories'. However, we still find that TICL correctly points out that 'necklaces' and 'anklets' should belong under 'jewelry'. This example again suggests that the taxonomy induction problem is a difficult problem, especially if we are looking to uncover very precise parent-child relationships.

Additionally, we quantitatively analyzed the performance of TICL taxonomies using four metrics: shortest-paths overlap, neighbors-overlap, modified tree edit distance, and inference accuracy-based metrics (precision, recall and F-measure). The shortest paths overlap and modified tree edit distance measure the 'global' quality of the induced taxonomy, while the other metrics capture the 'local' quality. More details about these four measurement metrics are included in [9]. In all cases, we find that the proposed TICL approach consistently delivers superior performance compared with other baseline methods such as MaxTransGraph [28], MaxTransForest [29], and TAXI [30].

4 System Workflow

In this section, we briefly describe the architecture and system workflow for InVInT. The workflow of InVInT was designed to be as seamless as possible for users by requiring a minimum of 'clicks' for the user to get started. This potentially allows the system to be executed over the Web, or even be offered as a cloud service, if so desired. Although InVInT comes with default values for many of its parameters,

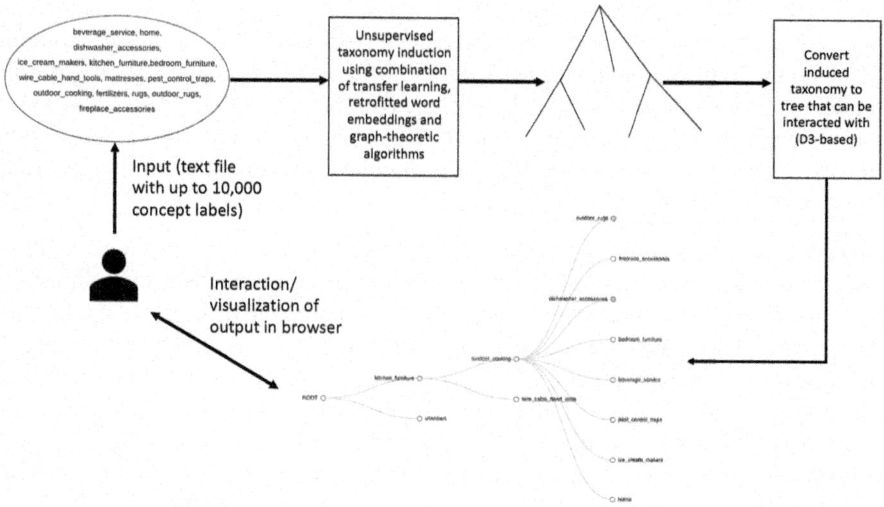

Fig. 2 A simple schematic of InVInT, given only a concept-set of e-commerce terms (e.g., products) by the user. The visualization opens in a browser, and is fast and interactive

more experienced or interested users are also given the option to customize options through either a web interface or the command line. InVInT was demonstrated in the 2021 ASONAM conference, where we showed how users had the ability to try the system out on their own concept-label sets, input simply as text files (described below).

Input As shown in Fig. 2, the input to the system is simply a set of *phrasal concept labels*. The concept labels are phrasal because they may not be single words (e.g., 'beverage_service'). There is no additional information assumed to be available about them.[4] As described earlier, the goal is to induce a tree-like taxonomy over these concepts and produce the taxonomy in a visual form that the user can interact with. While users could input their own concept sets (via a simple text file), an actual tutorial of InVInT provides example text files containing data from well-known publicly available examples, such as subsets of concepts from the Google Product Taxonomy.[5]

TICL Algorithm We run the dockerized TICL algorithm to induce a global taxonomy over the input concept labels. Users can specify the parameters used in the algorithm. For example, one can set the maximum degree of each node in the

[4] This is in contrast to entities in knowledge graphs, for which additional context can be found (e.g., from Wikidata or GeoNames [25, 34]) if the entity can be linked to a canonical knowledge base, or if the knowledge graph has been constructed already from a large natural language corpus.

[5] https://www.google.com/basepages/producttype/taxonomy.en-US.txt.

output taxonomy when running the docker image with a command. If no parameters are specified, we also include default parameters specifying the use of pre-training embedding in the taxonomy generation and setting the maximum degree of each node in the taxonomy as 6. As mentioned in Sect. 3, TICL is an algorithmic GTC approach that has already been shown to be significantly more competitive than several other taxonomy induction techniques, such as MaxTransGraph [28], MaxTransForest [29] and TAXI [30].

Interactive Visualization In the real world, concept-labels over which the taxonomy has to be induced can number in the hundreds or even thousands (as we showed in one of our recent papers [31]). For any induced taxonomy to be useful, the user must be able to interact with the taxonomy without getting overwhelmed. We developed the original technology assuming a general mix of both technical and non-technical users (e.g., content producers, designers and marketers). Therefore, the final output does not require complicated set up, and using D3 [35], can be visualized on a standard browser. It is also interactive; users can click on 'nodes' in the tree (specific concept-labels) to expand and see the set of child-nodes for that node.

In practice, this visualization is useful even for algorithmic developers, since they could tune various parameters and visualize a set of trees in different tabs on the browser to gain a more intuitive understanding of algorithmic performances and qualitative differences.

4.1 Evaluation of InVInT

To gain an intuitive sight of the performance of InVInT (which comprises both the TICL algorithm and visualization), we sampled a set of 15 concepts from each of five product branches (i.e., business and industrial supplies, clothing, health care, media and pet) derived from a real-world e-commerce taxonomy, and input each of these 5 concept-sets to both the TICL algorithm as well as a Random taxonomy induction baseline. Taxonomies induced by both methods were then fed into the visualization component of InVInT to generate taxonomies.

Note that the Random baseline is similar to TICL, but differs in the second step. Recall that TICL will find the k closest concepts of every single node to construct an approximately k-regular graph. The baseline also builds an approximately k-regular graph, but it chooses to randomly connect every single node to k other nodes.

Next, we hand over the visualization of the two generated taxonomies (for each category-set input) to a 'blind' reviewer. Similar to A/B testing, the review is bling because the source of generated taxonomies (Random or TICL) is kept hidden from the reviewer. The reviewer uses the label "Better" or "Worse" to indicate which of the pair of taxonomies is better. To enforce further objectivity, the annotation is done "offline" i.e., the individual who generated the taxonomies and named the files was not present in the room when the individual who received the files annotated them. Additionally, using the fifteen concepts per input-set, an 'ideal' taxonomy is also

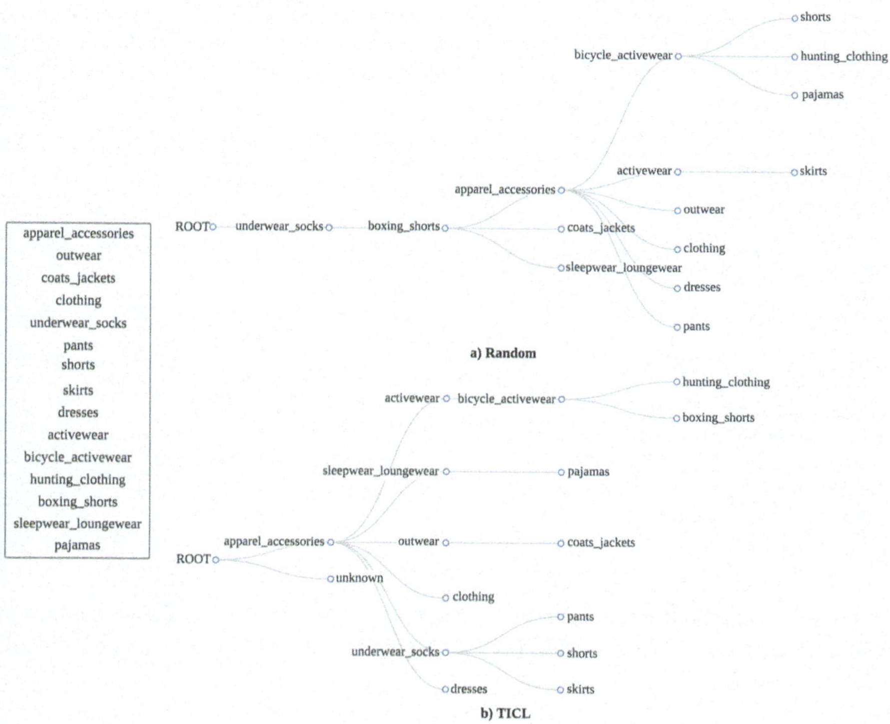

Fig. 3 The TICL and Random baseline results for an example set of 15 e-commerce terms derived using a real-world product taxonomy

provided by the same reviewer as a reference, which can be used to quantitatively (in absolute terms) measure the performance of both the underlying TICL algorithm and the Random baseline.

In all five cases, the reviewer picked out the TICL-generated taxonomy as a better taxonomy. By way of example, we provide a pair of generated taxonomies in Fig. 3 given *Clothing* terms as input. Based on the 'ideal' taxonomy provided by the reviewer, we use tree edit distance and the intersection of edge sets to measure the similarity between generated taxonomies and the 'ideal' taxonomy. In Table 1, the tree edit distance results show that the TICL algorithm achieves a much better performance in *Business and industrial supplies*, *Clothing*, and *Pet* test cases and a slightly better performance in *Media* and *Pet*, which are all consistent with the reviewer's annotation.

Besides, we find that the TICL-generated taxonomies also share more edges with the 'ideal' (recall that these were generated by the annotator) taxonomies than the Random-generated taxonomies in Table 2. For example, in the *Clothing* example shown in Fig. 3, the TICL-generated taxonomy shares 8 edges with the 'ideal' taxonomy, while the Random-generated one only shares 4 edges. On average, the TICL-generated taxonomy shares 2.4 more edges with the corresponding 'ideal'

Table 1 The tree edit distance between the generated taxonomies and the 'ideal' taxonomy in five test cases. A shorter tree edit distance indicates that two taxonomies are more similar. For ease of interpretation, we also include the relative performance improvement of TICL over the Random baseline

Category	Random	TICL	Relative improvement of TICL over random
Business_industrial	230	177	%
Clothing	167	104	%
Health_care	192	191	%
Media	247	243	%
Pet	194	119	%

Table 2 The edge overlap between the generated taxonomies and the 'ideal' taxonomy in five test cases. More edge overlap indicates that two taxonomies are more similar

Category	Random	TICL
Business_industrial	1	2
Clothing	4	8
Health_care	0	4
Media	3	6
Pet	4	4

taxonomy than the Random-generated taxonomy. The proportion of shared edges in the edge set is very high considering that there are a total of 16 edges in the 'ideal' taxonomy.

5 Discussion

Even with the use of advanced techniques, as incorporated in TICL, such as representation learning and retrofitting, global taxonomy construction (GTC) remains a hard problem. In turn, this illustrates the important role that visualization can play. Importantly, while a taxonomy may formally be deconstructed as a 'set of edges', users perceive the final output as a tree. By separating the taxonomy induction process from the visualization (as demonstrated through the preliminary annotation experiment whereby the same visualization component was used for the taxonomies induced by both the Random baseline and the TICL algorithm over the same concept-set), InVInT can also help users to compare and contrast different algorithms, especially when the edge-overlap or tree-edit distance metrics are not very interpretable, despite their greater quantifiability.

The example fragments shown in the figures also suggest that TICL (and InVInT) may be having a more difficult time determining the precise nature of the relationship between two semantic categories. For example, is one product a 'sub-product' of another, or is there some other overlap not easily explained by such parent-child relationships? For example, might the products be *complements*, in which case the customer would buy them together, or are they *alternatives* to one another? In the economics literature, such relationships have been widely studied, but automatically deducing them from raw strings remains difficult for automated processes to achieve. Once again, visualization can help the user to make sense of an algorithm's output. To the best of our knowledge, no GTC algorithm currently exists for determining such domain-specific relations automatically given only background knowledge and no training data.

5.1 Limitations and Future Work

Although InVInT is a promising attempt at the GTC problem, it is also subject to some limitations that we summarize below:

1. First, one of the limitations of InVInT is the requirement by TICL that background resources be available for retrofitting in the embedding space. This implies that, before the system can be set up and used in a new domain, a user would have to locate relevant background resources and set up TICL accordingly. Although this problem generally occurs in almost all systems when domain transfer is required, we hope to address it in future work (e.g., by integrating a Web scraper into the system, as discussed below).
2. Second, the quality of induced taxonomies can suffer for certain situations, as we illustrated earlier through Fig. 1. We hypothesize that there remain significantly opportunities for performance gain in this space, and InVInT is only intended to be an early prototype.
3. Finally, the retrofitting algorithm that is incorporated into TICL can be significantly improved due to advances in transformers and other kinds of neural networks. In a similar vein, graph neural networks and other such advanced architectures could be used in lieu of TICL's minimum spanning tree algorithm.

In future research, we hope to address these limitations and challenges by improving TICL and potentially integrating an automatic Web scraper into the architecture to look for background resources (given any concept-label set) that can be retrofitted in the embedding space. We will also consider upgrading the retrofitting algorithm itself, including with more advanced transformer-based algorithm that can be fine-tuned, to improve performance and robustness. An even more ambitious agenda would be to consider the incorporation of large-scale *generative* models such as GPT-3 to elicit knowledge that could be used to build high-quality global taxonomies.

6 Conclusion

Domain-specific global taxonomy construction (GTC) is a challenging problem with valuable real-world applications in domains, such as e-commerce. Given concepts represented as labels, and a limited set of background resources, the GTC problem involves inducing a taxonomy over the concept-set using the minimal inputs provided. We described the architecture and workflow of a system called InVInT (Interactive Visualization of Induced Taxonomies), which uses a state-of-the-art 'transfer-based' approach called TICL as its underlying GTC approach. Next, it uses open-source tools, such as D3, to interactively visualize the generated taxonomy in a browser. The system also allows users to upload their own concept-sets and is lightweight, with default values for most parameters. Advanced users, however, can customize the system with parameters of their own to achieve its full benefits. An early version of InVInT was developed in collaboration with researchers in industry, and the TICL algorithm, which is key to its functioning, has already been evaluated on challenging e-commerce datasets.

Acknowledgments The authors would like to thank Nicolas Torzec and Chien-Chun Ni from the Yahoo! Knowledge Graph Research Group for helping us with the data and visualization code, which is based on open-source packages. This project was primarily funded under a Yahoo! Faculty Research Engagement Program grant awarded to Kejriwal.

References

1. Kejriwal, M., Shen, K.: Unsupervised real-time induction and interactive visualization of taxonomies over domain-specific concepts. In: Proceedings of the 2021 IEEE/ACM International Conference on Advances in Social Networks Analysis and Mining, pp. 301–304 (2021)
2. Kim, Y.S.: Recommender system based on product taxonomy in e-commerce sites. J. Inf. Sci. Eng. **29**(1), 63–78(2013)
3. Davulcu, H., Koduri, S.,Nagarajan, S.: Datarover: a taxonomy based crawler for automated data extraction from data-intensive websites. In: Proceedings of the 5th ACM International Workshop on Web Information and Data Management, pp. 9–14 (2003)
4. PriceGrabber. http://www.pricegrabber.com/. Accessed 2017
5. Yu, J., Buyya, R.: A taxonomy of scientific workflow systems for grid computing. ACM Sigmod Rec. **34**(3), 44–49 (2005)
6. ICD: International Classification of Diseases 11th Revision. https://icd.who.int/en. Accessed 19 Aug 2022
7. Google: Google product category. https://support.google.com/merchants/answer/6324436?hl=en. Accessed Aug 2011
8. Uddin, M.N., Janecek, P.: Performance and usability testing of multidimensional taxonomy in web site search and navigation. Perform. Meas. Metrics **8**, 18–33 (2007)
9. Kejriwal, M., Shen, K., Ni, C.-C., Torzec, N.: Transfer-based taxonomy induction over concept labels. Eng. Appl. Artif. Intell. **108**, 104548 (2022)
10. Mikolov, T., Sutskever, I., Chen, K., Corrado, G.S., Dean, J.: Distributed representations of words and phrases and their compositionality. In: Proceedings of the 26th International Conference of Neural Information Processing Systems (NIPS' 13), vol. 2, pp. 3111–3119. Curran Associates, Red Hook (2013)
11. Devlin, J., Chang, M.-W., Lee, K., Toutanova, K.: Bert: Pre-training of deep bidirectional transformers for language understanding. Preprint (2018). arXiv:1810.04805
12. Pennington, J., Socher, R., Manning, C.D.: Glove: global vectors for word representation. In: Proceedings of the 2014 Conference on Empirical Methods in Natural Language Processing (EMNLP), pp. 1532–1543 (2014)

13. Joulin, A., Grave, E., Bojanowski, P., Mikolov, T.: Bag of tricks for efficient text classification. Preprint (2016). arXiv:1607.01759
14. Liu, Y., Ott, M., Goyal, N., Du, J., Joshi, M., Chen, D., Levy, O., Lewis, M., Zettlemoyer, L., Stoyanov, V.: Roberta: a robustly optimized bert pretraining approach. Preprint (2019). arXiv:1907.11692
15. Brown, T.B., Mann, B., Ryder, N., Subbiah, M., Kaplan, J., Dhariwal, P., Neelakantan, A., Shyam, P., Sastry, G., Askell, A., et al.: Language models are few-shot learners. Preprint (2020) . arXiv:2005.14165
16. Ma, T., Pan, Q., Rong, H., Qian, Y., Tian, Y., Al-Nabhan, N.: T-bertsum: Topic-aware text summarization based on bert. IEEE Trans. Comput. Soc. Syst. **9**(3), 879–890 (2021)
17. Hakala, K., Pyysalo, S.: Biomedical named entity recognition with multilingual bert. In: Proceedings of the 5th Workshop on BioNLP Open Shared Tasks, pp. 56–61 (2019)
18. Chang, Y., Kong, L., Jia, K., Meng, Q.: Chinese named entity recognition method based on bert. In: 2021 IEEE International Conference on Data Science and Computer Application (ICDSCA), pp. 294–299. IEEE (2021)
19. Liu, A., Huang, Z., Lu, H., Wang, X., Yuan, C.: Bb-kbqa: Bert-based knowledge base question answering. In: China National Conference on Chinese Computational Linguistics, pp. 81–92. Springer (2019)
20. Luo, D., Su, J., Yu, S.: A bert-based approach with relation-aware attention for knowledge base question answering. In: 2020 International Joint Conference on Neural Networks (IJCNN), pp. 1–8. IEEE (2020)
21. Perozzi, B., Al-Rfou, R., Skiena, S.: Deepwalk: online learning of social representations. In: Proceedings of the 20th ACM SIGKDD International Conference on Knowledge Discovery and Data Mining, pp. 701–710 (2014)
22. Grover, A., Leskovec, J.: node2vec: scalable feature learning for networks. In: Proceedings of the 22nd ACM SIGKDD International Conference on Knowledge Discovery and Data Mining, pp. 855–864 (2016)
23. Ristoski, P., Paulheim, H.: Rdf2vec: Rdf graph embeddings for data mining. In: Semantic Web Conference, pp. 498–514. Springer (2016)
24. Kejriwal, M.: Domain-Specific Knowledge Graph Construction. Springer, Berlin (2019)
25. Kejriwal, M., Szekely, P.: Neural embeddings for populated geonames locations. In: International Semantic Web Conference, pp. 139–146. Springer (2017)
26. Snow, R., Jurafsky, D., Ng, A.Y.: Semantic taxonomy induction from heterogenous evidence. In: Proceedings of the 21st International Conference on Computational Linguistics and the 44th Annual Meeting of the Association for Computational Linguistics, pp. 801–808. Association for Computational Linguistics (2006)
27. Velardi, P., Faralli, S., Navigli, R.: Ontolearn reloaded: a graph-based algorithm for taxonomy induction. Comput. Linguist. **39**(3), 665–707 (2013)
28. Berant, J., Dagan, I., Goldberger, J.: Learning entailment relations by global graph structure optimization. Comput. Linguist. **38**(1), 73–111 (2012). https://doi.org/10.1162/COLI_a_00085
29. Berant, J., Alon, N., Dagan, I., Goldberger, J.: Efficient global learning of entailment graphs. Comput. Linguist. **41**(2), 249–291 (2015). https://doi.org/10.1162/COLI_a_00220
30. Panchenko, A., Faralli, S., Ruppert, E., Remus, S., Naets, H., Fairon, C., Ponzetto, S.P., Biemann, C.: Taxi at semeval-2016 task 13: a taxonomy induction method based on lexico-syntactic patterns, substrings and focused crawling. In: Proceedings of the 10th International Workshop on Semantic Evaluation, San Diego, CA, USA. Association for Computational Linguistics (2016)
31. Kejriwal, M., Selvam, R.K., Ni, C.-C., Torzec, N.: Locally constructing product taxonomies from scratch using representation learning. In: Proceedings of the 12th IEEE/ACM International Conference on Advances in Social Networks Analysis and Mining (ASONAM '20), pp. 507–514. IEEE (2021). https://doi:.org/10.1109/ASONAM49781.2020.9381320
32. Faruqui, M., Dodge, J., Jauhar, S.K., Dyer, C., Hovy, E., Smith, N.A.: Retrofitting word vectors to semantic lexicons. Preprint (2014). arXiv:1411.4166
33. Kruskal, J.B.: On the shortest spanning subtree of a graph and the traveling salesman problem. Proc. Am. Math. Soc. **7**(1), 48–50 (1956)

34. Vrandečić, D., Krötzsch, M.: Wikidata: a free collaborative knowledgebase. Commun.ACM **57**(10), 78–85 (2014)
35. Bostock, M., Ogievetsky, V., Heer, J.: D^3 data-driven documents. IEEE Trans. Vis. Comput. Graph. **17**(12), 2301–2309 (2011)

Exploring Online Video Narratives and Networks Using VTracker

Thomas Marcoux, Oluwaseyi Adeliyi, Dayo Samuel Banjo, Mayor Inna Gurung, and Nitin Agarwal

Abstract YouTube is the second most popular website on the internet and a major actor in information propagation, therefore making it efficient as a potential vehicle of misinformation. Current tools available for video platforms tend to hyperfocus on metadata aggregation and neglect the analysis of the actual videos. In an attempt to provide analysts the tools they need to perform various research (behavioral, political analysis, sociology, etc.), we present VTracker (formerly YouTubeTracker), an online analytical tool. Some of the insight analysts can derive from this tool are inorganic behavior detection and algorithmic manipulation. We aim to make the analysis of YouTube content and user behavior accessible not only to information scientists but also communication researchers, journalists, sociologists, and many more. We demonstrate the utility of the tool through some real world data samples.

Keywords Youtube · Misinformation · Disinformation · VTracker · Bots

This research is funded in part by the U.S. National Science Foundation (OIA-1946391, OIA-1920920, IIS-1636933, ACI-1429160, and IIS-1110868), U.S. Office of Naval Research (N00014-10-1-0091, N00014-14-1-0489, N00014-15-P-1187, N00014-16-1-2016, N00014-16-1-2412, N00014-17-1-2675, N00014-17-1-2605, N68335-19-C-0359, N00014-19-1-2336, N68335-20-C-0540, N00014-21-1-2121, N00014-21-1-2765, N00014-22-1-2318), U.S. Air Force Research (FA9550-22-1-0332), U.S. Army Research Office (W911NF-20-1-0262, W911NF-16-1-0189), U.S. Defense Advanced Research Projects Agency (W31P4Q-17-C-0059), Arkansas Research Alliance, the Jerry L. Maulden/Entergy Endowment at the University of Arkansas at Little Rock, and the Australian Department of Defense Strategic Policy Grants Program (SPGP) (award number: 2020-106-094). Any opinions, findings, and conclusions or recommendations expressed in this material are those of the authors and do not necessarily reflect the views of the funding organizations. The researchers gratefully acknowledge the support.

T. Marcoux (✉) · O. Adeliyi · D. S. Banjo · M. I. Gurung · N. Agarwal
COSMOS Research Center, University of Arkansas at Little Rock, Little Rock, AR, USA
e-mail: txmarcoux@ualr.edu; odadeliyi@ualr.edu; sbanjo@ualr.edu; mgurung@ualr.edu; nxagarwal@ualr.edu
https://cosmos.ualr.edu

S. T. Özyer, B. Kaya (eds.), *Cyber Security and Social Media Applications*, Lecture Notes in Social Networks, https://doi.org/10.1007/978-3-031-33065-0_6

1 Introduction

Online video platforms can deliver a complex message in a way that captures the audience's attention more effectively than text-based platforms. Because of this, platforms like YouTube have become some of the most relevant in the age of digital mass communication. Efforts to influence opinions are omnipresent in this space and especially active during events which tend to polarize political opinions.

Although there exists many studies focusing on online social network misinformation, especially on popular platforms such as Twitter, we believe that video-based platforms are not receiving as much attention as they should. A few reasons for that are variations in privacy agreements and terms of service, data accessibility, but most of all, the nature of the data. There are many natural language processing tools ready to be leveraged to analyze text-based platforms. However, rich media such as videos do not share such systematic scientific methodologies. While some online tools exist to provide analysis of content engagement or estimated earning reports, we have not found tools attempting to tackle the issues of video summarization and characterization necessary to handle the big data issue of video platforms. In an attempt to provide analysts the tools they need to perform various research (behavioral, political analysis, sociology, etc.), we present VTracker (formerly YouTubeTracker). In the subsequent sections, we briefly highlight some of the state of the art technologies in YouTube analysis, and then discuss some of the features and capabilities of the VTracker application; along with real world cases, namely propaganda videos relevant to the Indo-Pacific Region, NATO Baltic Operations from 2020, and the 2020 Canada Elections.

This work extends a previous publication [18] by introducing new features available in the tool described here.

2 Literature Review

According to third party web traffic reports [24], YouTube is the second most popular website and accounts for 20.6% of search traffic. According to official YouTube sources [9], 1 billion hours of videos are watched each day. Another study by Cha et al. [4] found that 60% of YouTube videos are watched at least 10 times on the day they are posted. The authors also highlight that if a video does not attract viewership in the first few days after upload, it is unlikely to attract viewership later on. YouTube provides an overwhelming amount of streaming data: over 500 hours of videos are uploaded every minute on average. A number which was "only" 300 in 2013 [8]. In previous publications [10, 19] we identified YouTube as a potential vehicle of misinformation. We proposed the use of YouTube metadata for understanding and visualizing these phenomena by observing data trends. We also proposed the concept of movie barcodes as a tool for video summarization clustering [7]. In this publication, we present the movie barcode tool as a part of VTracker, as well as new video characterization tools. Previous research [5] has looked into engagement patterns of YouTube videos and highlighted the related videos engagement trends, later designated as the "rabbit hole effect" where users will be recommended increasingly relevant videos. In some cases, where the subject matter is a very polarizing one, this effect has been shown to be a contributing factor in user radicalization [22]. This last study takes the example of vaccine misinformation, which has attracted much interest from the information community. With some research highlighting that while users turn to YouTube for health information, many of the resources available failed to provide accurate information [3, 17], and public institutions should increase their online presence [6] to make reliable information more accessible. Other research also identified impacts on climate [16]. Recent research on the same subject leverages advanced NLP techniques on text entities such as video comments [20] but we could find little work available on the video content itself.

Researchers have examined community behavior, when it comes to moderation [13], or identifying influential groups within communities and their potential as vehicles of misinformation [2]. There are few studies where the video content itself is considered, but more recent works are starting to emerge and use modern NLP techniques to extract messages from video content [12, 21]. Many studies analyzing the impact of misinformation also tend to use smaller sets of videos and rely on either manual or self-reported results. Others take the forms of audits [11, 23], checking YouTube for biased behaviors and vulnerability to "rabbit hole effects". This results in one of the primary motivation behind this research: the lack of systemic, large scale tool analyzing YouTube content patterns and their role in the spread of misinformation.

3 VTracker

VTracker is a web application that provides valuable insights in a drilled down version from YouTube data. In this section, we describe some of the new and improved features and analytical capabilities of VTracker.

3.1 Data Collection

One possible reason for the lack of large scale YouTube research is the difficulty of data collection. Collection needs to be conducted through Google's official API, and comply with terms of service preventing the sharing of collected data, especially sharing identifying information. For this reason, little annotated data is available for research. Should researchers undergo YouTube data collection through their own means, the task is challenging and calls for a serious storage solution to house the volume of data required to unlock significant findings. This challenge is overcome here by recent research leveraging multiprocessing and relational databases, making YouTube data available for internal research [15].

3.2 *Movie Barcode Feature*

Figure 1 shows three videos belonging to a tracker focused on COVID-19 content. This is a high-level view of the tracker that shows its member videos' basic information, as well as one more advanced visualization: VTracker introduces a movie barcode feature aiming to detect embedded video clips sharing similar narratives. Representing videos as two dimensional image objects allows us to quickly detect identical video clips. In Fig. 1, this is represented by a narrow colored bar below each item. Clearly segmented barcodes can also indicate specific events and help cluster similar videos. Each video can be individually explored further by interacting with the video. This action opens a modal displaying further information about the video, as well as a detailed barcode allowing for navigating approximate events within the video.

3.3 *Emotion Analysis Feature*

Figure 2 shows a snippet of the content analysis page on VTracker, which focuses on insight around the video content, as opposed to engagement behaviors or posting

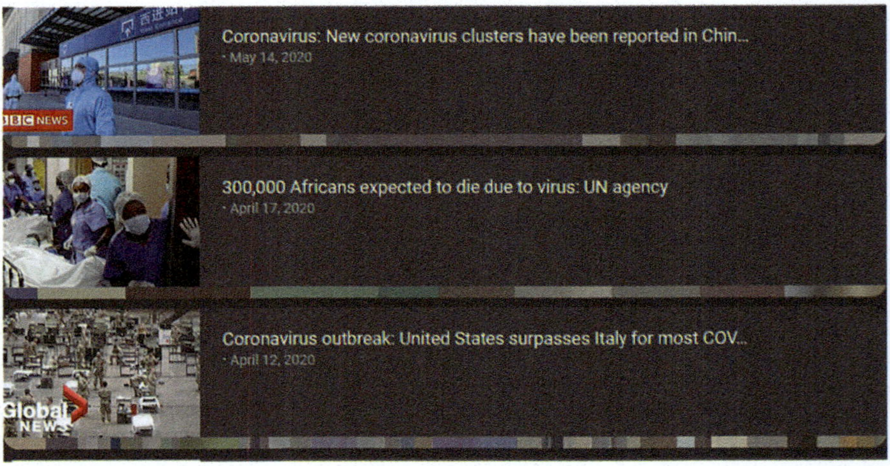

Fig. 1 Example of videos within a tracker showing movie barcodes

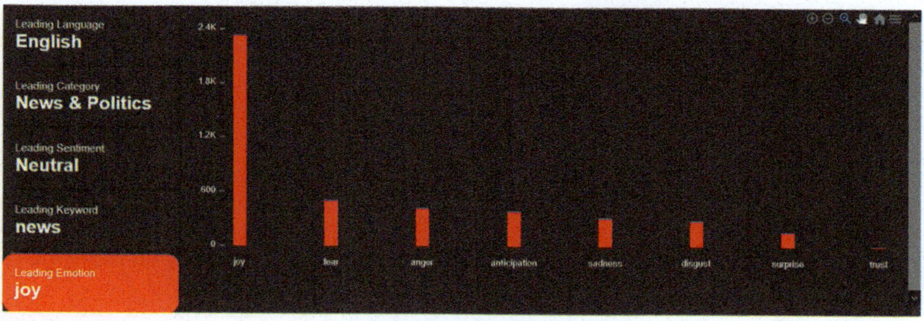

Fig. 2 Bar chart showing Leading Emotion for videos in a tracker

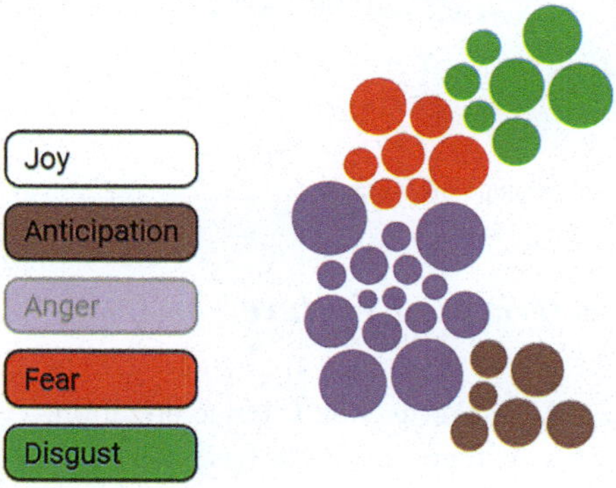

Fig. 3 Example of emotion filtering on the Cluster chart in VTracker

patterns. The snippet displays a bar chart representation of the leading emotion distribution within every video in a tracker, as computed by the public EmoNet library [1]. In Fig. 2, we can see that the leading emotion for this tracker seems to be joy. This is unexpected behavior given the nature of the tracker and we will experiment with other tools. The leading emotion for the videos is displayed alongside the leading language, leading category, leading opinion (approximated using sentiment analysis) and leading keyword. The leading emotion is also available as a cluster chart, as shown in Fig. 3. The labels on the left-hand side of the cluster chart not only acts as a legend but also allows users to selectively filter out specific emotions and focus on other emotions. We can see an example of this in Fig. 3 where the joy emotion is filtered out to highlight the distribution of other emotions within the videos.

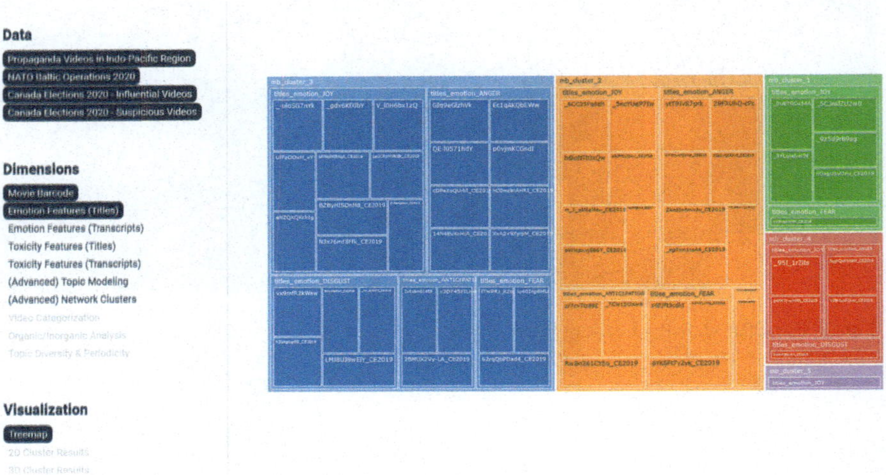

Fig. 4 Video characterization page on VTracker

3.4 *Video Characterization Feature*

Another new feature of VTracker is video characterization.[1] Video characterization uses various data points to categorize videos based on selected data and dimensions, using multiple views to visualize the relationship between these videos. This feature is useful in identifying patterns across videos with similar narratives, or using different features to identify clusters among a group of videos. Figure 4 shows the current implementation of this feature. It is divided into two main sections: the sidebar options on the left, and the chart display on the right. The sidebar includes, from top to bottom. what datasets to include, what features to include, and what visualization mode to use. The treemap uses colors to distinguish clusters, with a hierarchical view of the dimensions and the videos that fall under the selected dimensions. The data used in the characterization in Fig. 4 uses a sample of propaganda related to the Indo-Pacific Region, NATO Baltic Operations from 2020, influential videos and suspicious (showing inorganic behavior) videos from the 2020 Canada Elections. The data can be filtered by (de)selecting any of the data sets or features. The dimensions include the movie barcode (represented in clusters), emotion features(using titles and/or transcripts), toxicity features [14], topic modeling and network clusters. Some additional dimensions that will be added in future updates include the video categorization, (in)organic analysis, and topic diversity and periodicity. These dimensions can be selected or deselected by the user, causing the treemap to change based on the user's selection. For instance, by

[1] https://vtracker.host.ualr.edu/videoCharacterization.

interacting with the dimensions, we can see a more detailed level of these emotions as they relate to toxicity by selecting the Toxicity feature accordingly.

3.5 Network Feature

One of the features of VTracker is networking analysis (Fig. 5), which shows the visual representation of videos related to other videos. By clustering videos using a barcode, we could create a graph of videos as the node (vertex) and the degree of relationship as the edges between two videos. Using an undirected graph, the analysis shows each video as a node and the connection between two videos as an edge. This feature aims to visualize how connected a video content information is to another video. The videos are grouped into different clusters with barcode characterization. These clusters are related but still contain specific information and specialized key topics they focus on. The visualization is created by constructing a graph network of nodes with related information that has a connection to each other, with the use of content information generated from barcodes, we can characterize video clusters and visualize them using graphical network analysis. The same colors are used for all videos in a cluster on the graph network. The number of clusters is also shown, and the link represents if an edge exists between two videos. We gave an attribute of strength: the value of the edge in the graph between two nodes, determined by the degree of relation between two videos. We can make further assumptions with nodes that are not connected that convey different concentrations of the information from other videos even though they are related.

Figure 5 shows a snippet of the network analysis page on VTracker. The snippet displays a three-dimensional visualization of a video network of a tracker in the VTracker. The tracker has several videos about the Covid-19 extracted from youtube and extracted from this video title. The data used in the analysis relates the video of "Coronavirus: Lui Xiaoming, China's UK Ambassador", to the video of "Coronavirus: China dissatisfied". In the first video, Lui Xiaoming speaks on China's role in Covid-19, stating there is no guarantee that it was originated

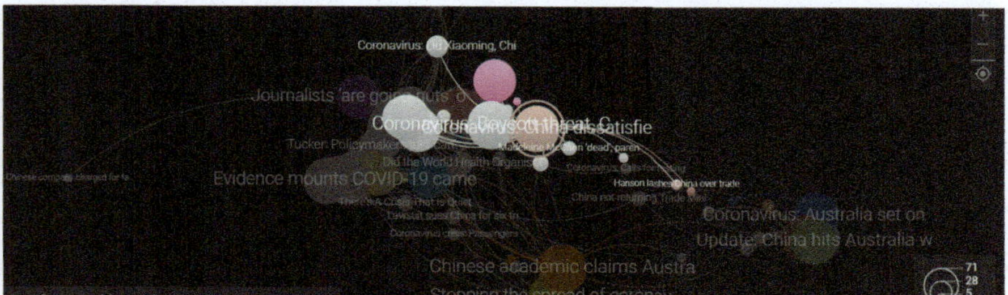

Fig. 5 Example of network graph

from China even though Wuhan, China recorded the first case. The second video expresses dissatisfaction over Australia's call for COVID investigation, which also focuses on how China's diplomacy of Covid-19 started. The videos corroborate each other and give a similar insight into the central topic. Different video clusters have similar colors, while videos in the same group have a similar color. This feature identifies patterns across the video in terms of the video content and narratives of the video. During the user interaction of the videos, a more detailed level of the associated videos is observed. The video network visualization offer scale-in and scale-out effect by the user to broadly and reduce the opacity of other nodes whose content description is less convey related information to the selected video.

Some of the challenges in this feature are the processing time and the CPU utilization of loading large numbers of videos and rendering them as a three-dimensional network graph. The network analysis computation time increases with the complexity of the graph and the number of different clusters formed.

3.6 *Narrative Visualization Feature*

The Narrative Analysis in VTracker, illustrated in Fig. 6, allows users to visually analyze narratives extracted by using the video and textual data. The text data includes the name of the channel, title, comments, and the description of the video. These data were explored by preprocessing via several natural language techniques such as tokenization, stop-word removal, and punctuation manipulation whereas,

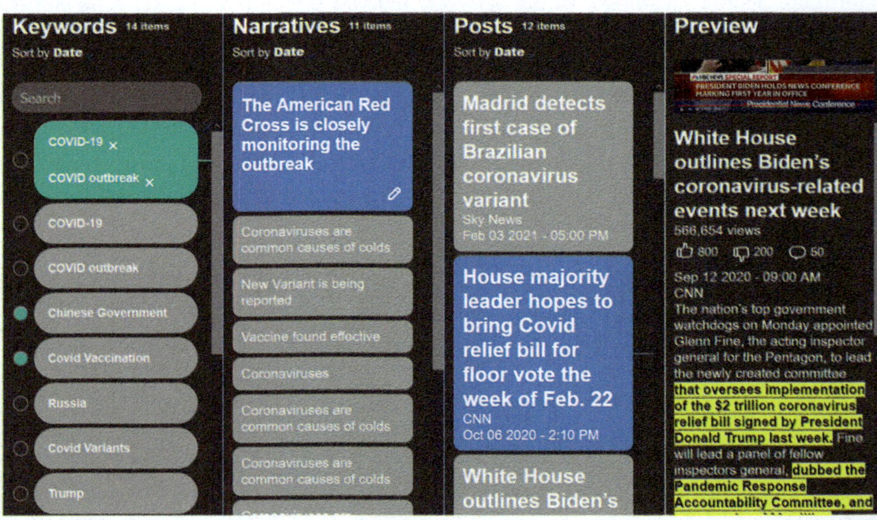

Fig. 6 Example of narrative tree demonstrating its different features

for the video data the transcripts were generated using multilevel processing sound extraction.

The visualization tool utilizes column layout arrangement, it enlists important entities in the keyword section and each of them has narratives related to it. The layout hides data complexity and enables users to focus on one keyword or group of keywords at a time. For example, if a user wants to see which narratives are associated with a particular keyword, they can choose that keyword and explore all the related narratives to it. Users also can explore narratives related to multiple overlapping keywords by grouping them together. For instance, in the given example different keywords relating to covid are grouped together. The grouped keywords can be ungrouped at once or each keyword can be removed individually. Furthermore, In the keyword column, the users also have the ability to search and add new keywords to be analyzed. The narratives are highly customizable as well. It gives the user the ability to edit the given narrative so that it can be reviewed by the moderators. The user-given feedback is utilized to improve the narrative extraction algorithm. Also, the tool is highly scalable as the extracted narratives are pre-calculated and sorted within the database. Once the user picks a narrative of interest the connected posts are given with information like the title of the video, the source, and published date which can be further expanded to visualize the preview of the video with user engagement information and the transcript of the video. Lastly, each column can be sorted with respect to the date, relevance, and alphabetical order. While the narrative tool allows users to analyze the narratives related to keywords of interest, it can be further expanded by adding the evolution of the narrative.

4 Future Works

In demonstrating this tool, we aim to bridge the gap between classical metadata analytics of online video platforms and the needed, more advanced, video data analysis. The VTracker tool attempts to meet this goal with some of the new features described in this paper, with more features being developed regularly. Some planned features include the following: Currently, only two visualization modes are available on the video characterization page: tree map, and node cluster chart. Other visualizations will be added moving forward to give the user more options to explore the data and reveal different patterns. More dimensions will also be integrated, as well as increased flexibility in terms of what data can be selected from.

Furthermore, we will introduce more advanced methods of detecting inorganic activity. For instance, by using rolling window correlation analysis. We are grouping data into rolling windows and computed pairwise correlation between the number of views, subscribers, total comments, and all published videos for each window. This was done to capture inauthentic behaviors such as a channel with decreasing subscribers but increasing views. From the arrays obtained and visualized (as seen in Fig. 7), we apply peak detection to isolate suspicious activities.

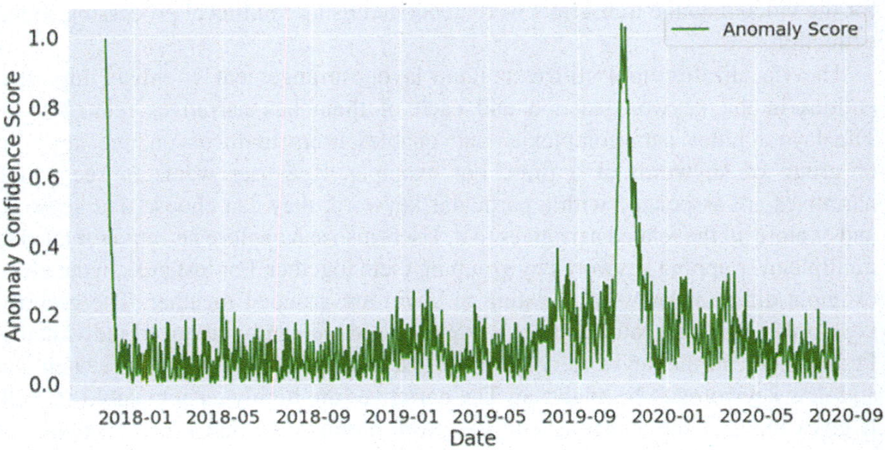

Fig. 7 Example of channel anomalies for views vs. videos (right)

Other planned features include: using these findings to create rule-based classification algorithm to compute and display a general suspicion score for each indicator, using co-commenter network behaviors to detect suspicious cliques of commenters that may act as artificial engagement boosts. This will let us detect inorganic clusters in a cross-channel capacity. We plan to integrate these features within VT as gauge-type indicators for each feature, as well as network graphs as is already implemented in Fig. 5.

Finally, a single contiguous movie barcode that allows users to navigate across all movie barcodes in a tracker will also be integrated. This will allow users to navigate video sets per features, for example dominant colors which would give indications of the video location.

References

1. Abdul-Mageed, M., Ungar, L.: EmoNet: fine-grained emotion detection with gated recurrent neural networks. In: Proceedings of the 55th Annual Meeting of the Association for Computational Linguistics (Volume 1: Long Papers), pp. 718–728. Association for Computational Linguistics, Vancouver (Jul 2017). https://doi.org/10.18653/v1/P17-1067, https://aclanthology.org/P17-1067
2. Alassad, M., Agarwal, N., Hussain, M.N.: Examining intensive groups in YouTube commenter networks. In: SBP-BRiMS (2019)
3. Basch, C.H., Hillyer, G.C., Meleo-Erwin, Z.C., Jaime, C., Mohlman, J., Basch, C.E.: Preventive behaviors conveyed on YouTube to mitigate transmission of COVID-19: cross-sectional study. JMIR Public Health Surveill. **6**(2), e18807 (2020). https://doi.org/10.2196/18807. http://www.ncbi.nlm.nih.gov/pubmed/32240096

4. Cha, M., Kwak, H., Rodriguez, P., Ahn, Y., Moon, S.: Analyzing the video popularity characteristics of large-scale user generated content systems. IEEE/ACM Trans. Netw. **17**(5), 1357–1370 (2009). https://doi.org/10.1109/TNET.2008.2011358
5. Cheng, X., Dale, C., Liu, J.: Statistics and social network of YouTube videos. In: 2008 16th International Workshop on Quality of Service, pp. 229–238 (2008). https://doi.org/10.1109/IWQOS.2008.32
6. Donzelli, G., Palomba, G., Federigi, I., Aquino, F., Cioni, L., Verani, M., Carducci, A., Lopalco, P.: Misinformation on vaccination: a quantitative analysis of YouTube videos. Human Vaccines Immunother. **14**(7), 1654–1659 (2018). https://doi.org/10.1080/21645515.2018.1454572, https://doi.org/10.1080/21645515.2018.1454572, publisher: Taylor & Francis _eprint: https://doi.org/10.1080/21645515.2018.1454572
7. Erol, R., Rejeleene, R., Young, R., Marcoux, T., Hussain, M.N., Agarwal, N.: YouTube video categorization using moviebarcode. In: The Sixth International Conference on Human and Social Analytics (HUSO 2020), Porto (2020)
8. Hale, J.: More than 500 hours Of content are now being uploaded to YouTube every minute (May 2019). https://www.tubefilter.com/2019/05/07/number-hours-video-uploaded-to-youtube-per-minute/
9. How YouTube Works - Product Features, Responsibility, & Impact. https://www.youtube.com/intl/en-GB/howyoutubeworks/
10. Hussain, M.N., Tokdemir, S., Agarwal, N., Al-khateeb, S.: Analyzing disinformation and crowd manipulation tactics on YouTube. In: Proceedings of the 2018 IEEE/ACM International Conference on Advances in Social Networks Analysis and Mining, ASONAM '18, pp. 1092–1095. IEEE Press (2018), event-place: Barcelona, Spain
11. Hussein, E., Juneja, P., Mitra, T.: Measuring misinformation in video search platforms: an audit study on YouTube. Proc. ACM Hum.-Comput. Interact. **4**(CSCW1) (2020). https://doi.org/10.1145/3392854, https://doi.org/10.1145/3392854. Association for Computing Machinery, New York
12. Jagtap, R., Kumar, A., Goel, R., Sharma, S., Sharma, R., George, C.P.: Misinformation detection on YouTube using video captions (2021). https://doi.org/10.48550/ARXIV.2107.00941, https://arxiv.org/abs/2107.00941
13. Jiang, S., Robertson, R.E., Wilson, C.: Bias misperceived:the role of partisanship and misinformation in YouTube comment moderation. In: Proceedings of the International AAAI Conference on Web and Social Media, vol. 13(01), pp. 278–289 (Jul 2019). https://ojs.aaai.org/index.php/ICWSM/article/view/3229
14. Jigsaw, Google: Perspective API (2021). https://perspectiveapi.com/
15. Kready, J., Shimray, S.A., Hussain, M.N., Agarwal, N.: YouTube data collection using parallel processing. In: 2020 IEEE International Parallel and Distributed Processing Symposium Workshops (IPDPSW), pp. 1119–1122 (2020). https://doi.org/10.1109/IPDPSW50202.2020.00185
16. Lemos, A.L.M., Bitencourt, E.C., Santos, J.G.B.d.: Fake news as fake politics: the digital materialities of YouTube misinformation videos about Brazilian oil spill catastrophe. Media Cult. Soc. **43**(5), 886–905 (2021). https://doi.org/10.1177/0163443720977301, https://doi.org/10.1177/0163443720977301
17. Li, H.O.Y., Bailey, A., Huynh, D., Chan, J.: YouTube as a source of information on COVID-19: a pandemic of misinformation? BMJ Global Health **5**(5), e002604 (2020). https://doi.org/10.1136/bmjgh-2020-002604, http://gh.bmj.com/content/5/5/e002604.abstract
18. Marcoux, T., Adeliyi, O., Agarwal, N.: Characterizing video-based online information environment using vtracker. In: Proceedings of the 2021 IEEE/ACM International Conference on Advances in Social Networks Analysis and Mining. pp. 297–300. ASONAM '21, Association for Computing Machinery, New York (2021). https://doi.org/10.1145/3487351.3489480, https://doi.org/10.1145/3487351.3489480

19. Marcoux, T., Agarwal, N., Adewale, O., Hussain, M.N., Galeano, K.K., Al-Khateeb, S.: Understanding information operations using YouTubeTracker. In: IEEE/WIC/ACM International Conference on Web Intelligence - Companion Volume, pp. 309–313. WI '19 Companion. Association for Computing Machinery, New York (2019). https://doi.org/10.1145/3358695.3360917, https://doi.org/10.1145/3358695.3360917. Event-place: Thessaloniki, Greece
20. Medina Serrano, J.C., Papakyriakopoulos, O., Hegelich, S.: NLP-based feature extraction for the detection of COVID-19 misinformation videos on YouTube. In: Proceedings of the 1st Workshop on NLP for COVID-19 at ACL 2020. Association for Computational Linguistics, Online (Jul 2020), https://www.aclweb.org/anthology/2020.nlpcovid19-acl.17
21. Stappen, L., Baird, A., Cambria, E., Schuller, B.W.: Sentiment analysis and topic recognition in video transcriptions. IEEE Intell. Syst. **36**(2), 88–95 (2021). https://doi.org/10.1109/MIS.2021.3062200
22. Tang, L., Fujimoto, K., Amith, M.T., Cunningham, R., Costantini, R.A., York, F., Xiong, G., Boom, J.A., Tao, C.: "Down the Rabbit Hole" of vaccine misinformation on YouTube: network exposure study. J. Med. Internet Res. **23**(1), e23262 (2021). https://doi.org/10.2196/23262, http://www.ncbi.nlm.nih.gov/pubmed/33399543
23. Tomlein, M., Pecher, B., Simko, J., Srba, I., Moro, R., Stefancova, E., Kompan, M., Hrckova, A., Podrouzek, J., Bielikova, M.: An audit of misinformation filter bubbles on YouTube: bubble bursting and recent behavior changes. In: Fifteenth ACM Conference on Recommender Systems, pp. 1–11. Association for Computing Machinery, New York (2021). https://doi.org/10.1145/3460231.3474241
24. Youtube.com Traffic Analytics and Market Share | Similarweb. https://www.similarweb.com/website/youtube.com/#overview

Twitter Credibility Score for Preventing Fake News Dissemination on Twitter

Hamza Taher and Reda Alhajj

Abstract Credibility on social media is hard to accomplish and this could affect the fate of nations like in the case of the 2016 U.S presidential elections (Bessi A, Ferrara E. First Monday 21(11), 2016), which makes fake news detection and prevention an attractive topic for research and development, especially when it comes to fighting the spread of fake news on Twitter as it is considered one of the biggest social media websites with an approximation of 330 million monthly active users all over the world (J. C. Twitter: number of monthly active users 2010–2019, 2019) and also considered one of the social websites that fake news spread violently and widely on (John S, James L. Knight Foundation. Disinformation, and influence campaigns on Twitter, 2018). This paper is an attempt to prevent fake news dissemination on Twitter by detecting deceptive tweets and generating a credibility score for Twitter accounts. Based on the assumption that deception style changes based on the topic we were able to increase the accuracy of style based deception detection to 65% and following the methodology presented in this paper we can decide the credibility score of a twitter account based on the style of writing. We applied this methodology on three different twitter accounts that belong to 44th, 45th, 46th presidents of the United States and we were able to calculate the credibility score for each account; moreover this methodology is applicable to any twitter account that has a sufficient number of tweets.

Keywords Fake news detection · Credibility · Fake news classification · Deception detection · Machine learning · Classification

---

H. Taher · R. Alhajj (✉)
Department of Computer Science, University of Calgary, Calgary, AB, Canada

Department of Computer Engineering, Istanbul Medipol University, Istanbul, Turkey

Department of Heath Informatics, University of Southern Denmark, Odense, Denmark
e-mail: ralhajj@medipol.edu.tr

S. T. Özyer, B. Kaya (eds.), *Cyber Security and Social Media Applications*, Lecture Notes in Social Networks, https://doi.org/10.1007/978-3-031-33065-0_7

6 Conclusion

Finally, we think that the results of this paper showed a lot of promise in the methodology used and proved our initial assumption on the changing behavior of the writing style of fake news. We hope that this research will be of help to future research and help increase the integrity and credibility of social media websites and especially Twitter by preventing the spread of misinformation.

7 Plan for Future Studies

This research is open for further improvements including finding better classification algorithms, choosing to detect fake news by other means such as the propagation-based fake news detection of each tweet, and increasing accuracy of the credibility score by considering other aspects like the credibility scores of other followers and following users to this account and the interaction between the accounts in the network.

References

1. G. Pennycook, J. McPhetres, Y. Zhang, J.G. Lu, D.G. Rand, Fighting COVID-19 misinformation on social media: Experimental evidence for a scalable accuracy-nudge intervention. Psychol. Sci. **31**(7), 770–780 (2020)
2. Roth, Y., Pickles, N.: Updating our approach to misleading information. Monday (11 May 2020)
3. X. Zhou, R. Zafarani, A survey of fake news: Fundamental theories, detection methods, and opportunities. ACM Comput. Surv. **53**(5), 1–40 (2020)
4. Wang, W.Y.: Liar, liar pants on fire: A new benchmark dataset for fake news detection. arXiv preprint arXiv:1705.00648 (2017)
5. Potthast, M., Kiesel, J., Reinartz, K., Bevendorff, J., Stein, B.: A stylometric inquiry into hyperpartisan and fake news. arXiv preprint arXiv:1702.05638 (2017)
6. Mitra, T., Gilbert, E.: Credbank: A largescale social media corpus with associated credibility annotations. In ICWSM'15 (n.d.)
7. K. Shu, A. Sliva, S. Wang, J. Tang, H. Liu, Fake news detection on social media: A data mining perspective. ACM SIGKDD Expl Newsl **19**(1), 22–369 (2017)
8. Davis, C.A., et al.: Botornot: A system to evaluate social bots. In: Proceedings of the 25th International Conference Companion on World Wide Web (2016)
9. F. Pedregosa et al., Scikit-learn: Machine learning in python. J Mach Learn Res **12**(85), 2825–2830 (2011)
10. L.E. Boehm, The validity effect: A search for mediating variables. Personal. Soc. Psychol. Bull. **20**(3), 285–293 (1994)
11. T. Kuran, C.R. Sunstein, Availability cascades and risk regulation. Stanford Law Rev. **1999**, 683–768 (1999)
12. R.S. Nickerson, Confirmation bias: A ubiquitous phenomenon in many guises. Rev. Gen. Psychol. **2**(2), 175–220 (1998)

1 Introduction

In a study by Pennycook et al. [1] they test an intervention intended to increase the truthfulness of the content people share on social media. They identify that sharing behavior improves after nudging the users to think about news accuracy. But the problem is that Social media websites do not want to implement a feature like that because sharing more content by the users is beneficial for them. Also, users do not want to get nudged (notified) every time they try to share a piece of information. This leads to the problem of how and when we should notify the user to check the information they share before trying to share it.

Twitter already updated their approach towards misleading tweets to limit their dissemination [2] by introducing new labels and warning messages that are manually generated which will provide additional context and information on some Tweets containing disputed or misleading information (Table 1).

This paper introduces a new approach by calculating the credibility of an account on social media by measuring the deceptiveness of the written style of the account's previous posts and giving a total credibility score of the account. This approach marks users that are involved in writing and disseminating deceptive news which can help to eliminate the problem of fake news. We apply this method on Twitter accounts, but it is also applicable to any other similar social media website.

2 Related Work

Many publications tried to solve the problem of fake news on the internet and especially on social media but there is no one unique approach that completely solves the problem of fake news. There are also some publications dedicated to solving the problem of fake news on Twitter by identifying bots like [8].

Table 1 Twitter labels and actions regarding misleading information, disputed, and unverified claims

Propensity for Harm	Moderate	Severe
Type	**Action Taken**	
Misleading Information	Label	Removal
Disputed Claim	Label	Warning
Unverified Claim	No action	No action

Twitter is a famous social media website that is easy to disseminate fake news on it due to multiple fundamental theories in Social Sciences mentioned in [3] which most of them apply to Twitter including:

- *Validity effect* [10]: Individuals tend to believe information is correct after repeated exposures.
- *Availability cascade* [11]: Individuals tend to adopt insights expressed by others when such insights are gaining more popularity within their social circles.
- *Confirmation bias* [12]: Individuals tend to trust information that confirms their preexisting beliefs or hypotheses.
- *Desirability bias* [13]: Individuals are inclined to accept information that pleases them.
- *Naïve realism* [14]: The senses provide us with direct awareness of objects as they are.

All of these are fundamental theories that social media websites like Twitter and Facebook benefit from because they keep the users engaged, connected, and interested.

Social media websites witnessed fake news dissemination which had tremendously negative effects on real-life such as using social media in the 2016 U.S. presidential election campaign where fake news stories outperformed real news on Facebook [15]. Twitter also had its fair share of falsehoods and rumors dissemination. By studying rumors dissemination on Twitter [16] found that rumors on Twitter spread faster and wider than the truth.

There is no agreed-upon definition in the scientific community for fake news until the moment of writing this paper but there is a common definition for it, fake news is a news article that is intentionally and verifiably false [7], intentionally means that it is created with the intention of misleading news consumers, verifiably means that the news article contains information that can be verified as false.

Detecting fake news on social media has been widely studied and researched due to how important it is to make social media a safe and credible place. There are multiple ways discussed in the scientific community to detect fake knows [3, 7] which are the following:

Knowledge-based fact-checking require checking the accuracy of the information that is included in the news, this is done in two ways:

- *Manual fact-checking* requires experts in the field of the information presented in the news to try to check the accuracy of the information based on their knowledge and their research or normal users.
- *Automated fact-checking* which relies on building knowledge graphs in which entities (subjects and objects) are represented nodes and relation (predicates) as edges. By Identifying and extracting entities and relations from the news and comparing them in the knowledge graphs we obtain the truth but the problem with this approach is that it requires heavy maintenance of the knowledge graphs because information and knowledge are always changing over time.

Style-based: which focuses on analyzing text features and finding the intention and the writing style of a news article using extracted statistical features based on the assumption that malicious News that is aimed to mislead the public has a writing style different from normal news. The extracted features based on the writing styles of the news can be:

- *General Features*: features describe content style from lexicon, syntax, discourse, and semantic features.
- *Latent Features*: where vectors represent words, sentences, and even documents in a news article and can be directly used as input to a classification model.

There are recent studies that have revealed features that differentiate between deceptive and honest news [16, 17]:

- *Informality*: Fake news is more informal (more swear words).
- *Diversity*: Fake news is more diverse (more unique verbs).
- *Subjectivity*: Fake news is less subjective (fewer report verbs).
- *Emotion*: Fake news is more emotional (more emotional words).

In this paper, we will use Style-based deceptive detection using latent features as input to our classification model to be able to detect deceptive news on Twitter.

Propagation-based This approach uses the network features and the discrimination patterns of news on social media to extract features that mark fake news from normal news. These features are:

- *Cascade Size*: Overall number of nodes in a cascade.
- *Cascade Breath*: Maximum and average breadth of a news cascade.
- *Cascade Depth*: Depth of a news cascade
- *Structural Virality*: Average distance among all pairs of nodes in a cascade.
- *Node Degree*: Degree of the root node of a news cascade.
- *Spread Speed*: Time taken for a cascade to reach a certain depth (or size).
- *Cascade Similarity*: Similarity scores between a cascade and other cascades.

Also, some recent studies pointed out some patterns that can differentiate fake news from normal news [18, 19] which are:

- *Distance*: Fake news Travels farther than normal news.
- *Spreaders*: Fake news has more nodes that spread the news
- *Engagement*: Fake news has stronger engagements.
- *Density*: Fake news propagation networks are more densely connected.

Although propagation-based fake news detection can be more robust than style-based detection, the propagation-based fake news needs a lot of time and data to be able to classify the news which makes this style of fake news detection on social media inefficient when it comes to early detection of fake news.

Source-based This style of fake news detection depends on the source of the news which is the first one to post the news. By assessing source credibility based on news authors and publishers or news consumers and propagators we can decide whether the news that they are writing is credible or not. This is done by analyzing the patterns and behavior of fake news authors and publishers. Also, there is some evidence that fake news propagators form networks and collaborate [19, 20] which is important when it comes to assessing the credibility of a specific author.

There are three main sources of fake news on the internet [7], defining and identifying these three main sources is critical to the fake news detection problem:

- *Social bots* are computer programs that can become malicious by designing them to purposely manipulate public opinion or help spread fake news or start a hashtag or a trend on social media, these bots can be controlled by a single person or and organization and pose a real threat on the credibility of the social media and any news that is shared on it.
- *Trolls* are the humans who spread fake news and aim the public opinion to achieve a personal goal and they are much harder to detect because unlike the social bots they don't have a specific pattern and therefore can't be easily classified.
- *Cyborgs* are the users that use bots and their input to efficiently spread fake news and they are the type that poses the most threat as they have both the benefits of bots (they use automated programs to spread fake news) and trolls.

3 Methodology

Our method is divided into different steps each with a specific purpose and an output that feeds the next steps.

Gathering Data: we start by gathering a corpora of tweets and news headlines that are already fact-checked from datasets like LIAR [4], BuzzFeed [5], and CREDBANK [6] so that we have different tweets on different topics. The more the data we collect in this step the more topics that we will discover and cluster and the better the performance of the topic classifier in the next steps.

After checking each dataset we decided to proceed with LIAR dataset which consists of 1256 rows of not only labels fake and true news but also has tags to indicate the topic of the news which helped us tremendously in the next steps when deciding the best clustering algorithm to adopt in our paper as having a labeled dataset can help us figure out the accuracy of the clustering algorithm.

Importing and Processing the Dataset: we imported the dataset as a python dataframe and dropped the unwanted columns [News ID, Subject(s), Speaker, State Info, Party Affiliation, Credit History] and kept the important columns which are [Label, News, Context].

For the Label column, the dataset labels the news based on six categories (True, Mostly True, Half True, Barely True, Fales, Pants on Fire).

We decided that true and mostly true are only the two labels which we will consider as truthful and marked them 1. And the rest will be considered as deceptive news and marked 0.

To ensure maximum efficiency and accuracy of the clustering algorithm we processed the news column using the NLTK library by extracting all the statements from the dataset and sanitizing each one by:

- Removing punctuation
- Tokenizing each statement into multiple words
- Removing stop words
- Stemming each word

For the context column, the dataset had multiple keywords for each statement with a total count of 134 keywords. To accurately predict the clustering algorithm performance, we had to choose only one keyword to represent the context of the news and created a new column named topic label which consists of 118 labels.

Evaluating Clustering Algorithms: in this step, we used the scikit-learn library [9] which provides many accurately implemented clustering algorithms from which we choose four well-known algorithms KMeans, Agglomerative, Birch, and Affinity algorithms.

To choose the best algorithm for clustering our dataset of news we decided to test the accuracy of each of them when the number of clusters is equal to the number of topic labels which is set to 118.

By using the statistical approach tf-idf to evaluate how relevant the news to each other and feeding the tf-idf matrix to the clustering models and then calculating the accuracy of each algorithm using the predetermined labels we obtained the clusters and accuracy results in (Table 2).

The highest accuracy we obtained is 21.423% from the Birch algorithm followed by the Agglomerative algorithm with an accuracy of 21.344%, however when considering other metrics that we used which are Silhouette, Calinski Harabasz, and Davies Bouldin scores we found that agglomerative algorithm defined the clusters better than Birch with nearly the same accuracy. Therefore, we decided to proceed with Agglomerative Clustering Algorithm (Figs. 1, 2, 3 and 4).

Choosing the Best Number of Clusters to decide what is the optimum number of clusters **K** to use in agglomerative clustering algorithm we tried to set **K** = **[5–20]** and calculated the accuracy, Silhouette, Calinski Harabasz, and Davies Bouldin scores of each **K** and obtained (Table 3).

Table 2 The results of the clustering algorithms

N-Clusters = N-Topics = 118				
	Agglo merative	KMeans	Birch	Affinity
Accuracy	21.344	18.261	21.42 3	18.814
Silhouette	0.071	0.021	0.016	0.011
Calinski Harabasz	5.038	1.964	2.208	3.289
Davies Bouldin	2.756	3.633	3.894	2.544

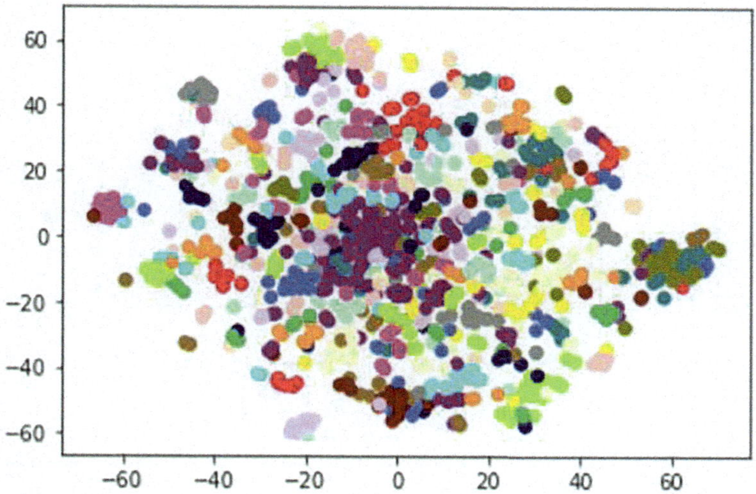

Fig. 1 Agglomerative clustering visualization

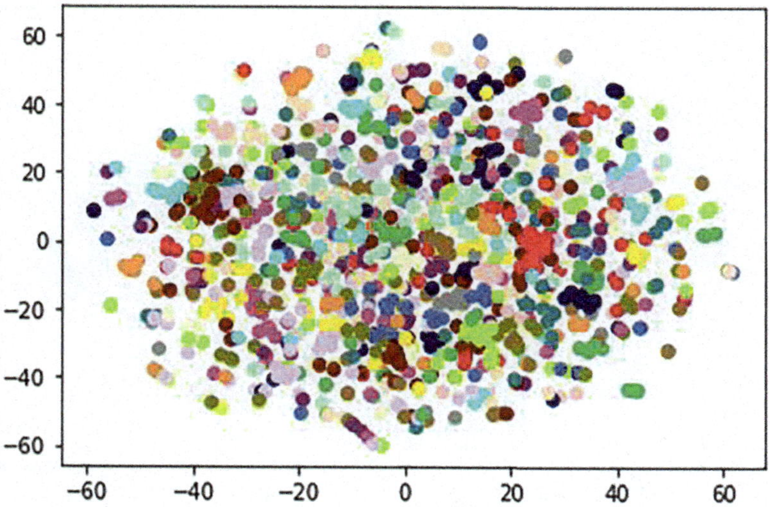

Fig. 2 KMeans clustering visualization

From (Figs. 5, 6, 7, and 8) we can see that the most interesting results are produced when **K** = (5, 6, 7, 8). At first, it might seem that when **K** = 5 is the best accuracy and best Davies Bouldin score so that is what we must choose but in fact both the accuracy and Calinski score have a negative correlation with **K**. The larger the K the smaller the accuracy and the Calinski score. Therefore, we focused on Silhouette and Davies Bouldin scores when Choosing the optimum number of clusters **K**. By looking at the Silhouette score and Davies score graphs we find that

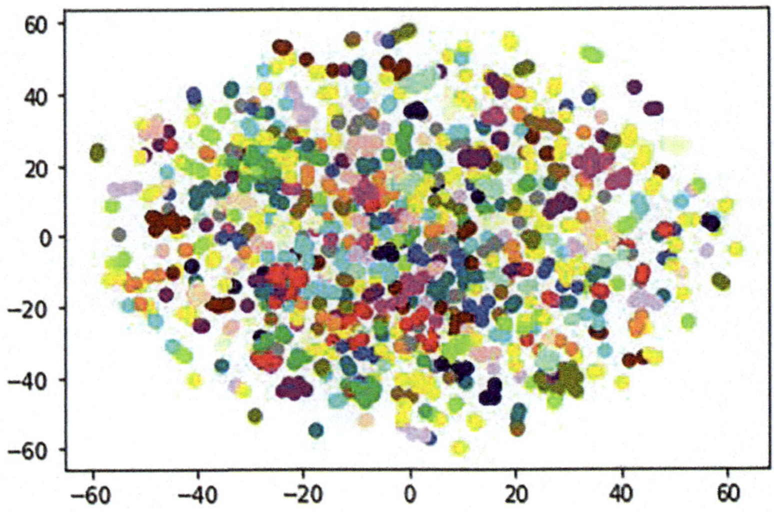

Fig. 3 Birch clustering visualization

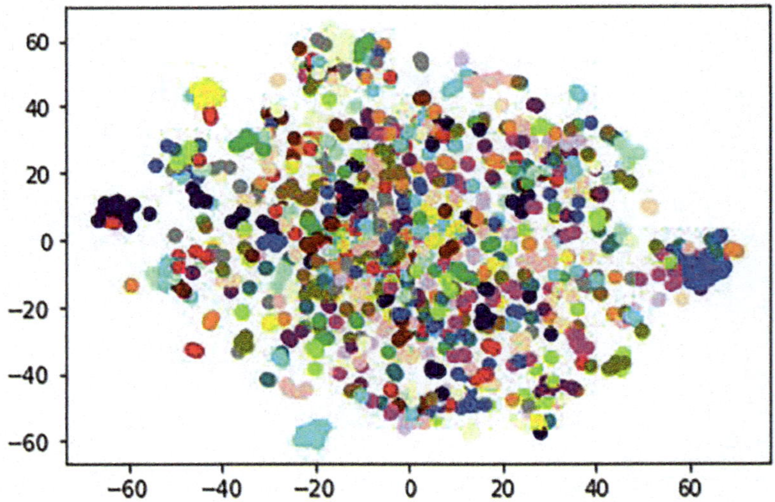

Fig. 4 Affinity clustering visualization

they both correlate when K = 8. Also, the accuracy at K = 8 is 66.798 which is above our acceptance criteria for the news clustering. That is why we decided to choose K = 8 to be the optimum number of news clusters.

Extracting Clusters after deciding the optimum number of clusters that we will use we performed agglomerative clustering one last time to extract each cluster news into a file. The final result of the clustering step is 8 files from cluster0 to cluster7

Table 3 The results of Agglomerative clustering algorithm

Scores With Different K clusters				
	K=5	K=6	K=7	K=8
Accuracy	69.565	68.142	67.984	66.798
Silhouette	0.073	0.076	0.075	0.076
Calinski Harabasz	21.179	19.647	18.539	17.680
Davies Bouldin	4.627	4.218	4.002	3.943

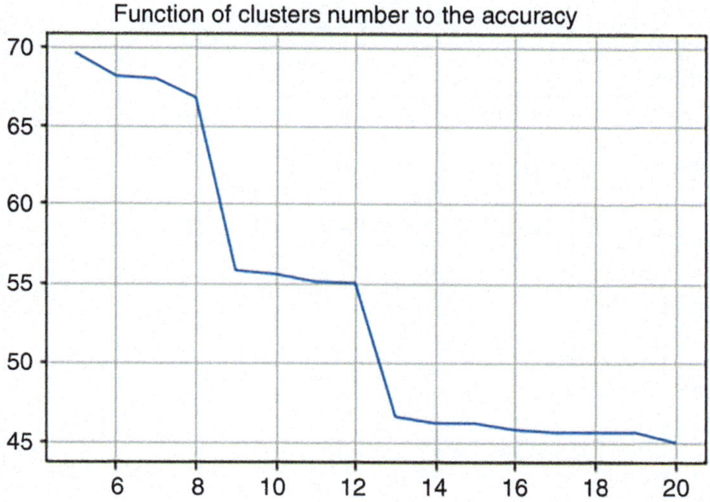

Fig. 5 Accuracy graph

each obtains the news statements that we are going to use to generate keywords that define cluster topics in the next step.

Generating Keywords in this step, we will use topic modeling to get an understanding of each topic and generate keywords that are specific to each news cluster we gathered. In this step we used sckit-learn as it provides a reliable implementation of the LDA (Latent Dirichlet Allocation) algorithm to automatically analyze each cluster we produced in the previous step and detect topics and extract keywords that indicate the topics so that we can classify new news using the extracted keywords in the next steps. By setting the number of wanted topics to be 10 and extracting the top 10 keywords we were able to identify each cluster and select the keywords that represent them from the top 100 words generated. Table 4 contains all the keywords extracted from the topics.

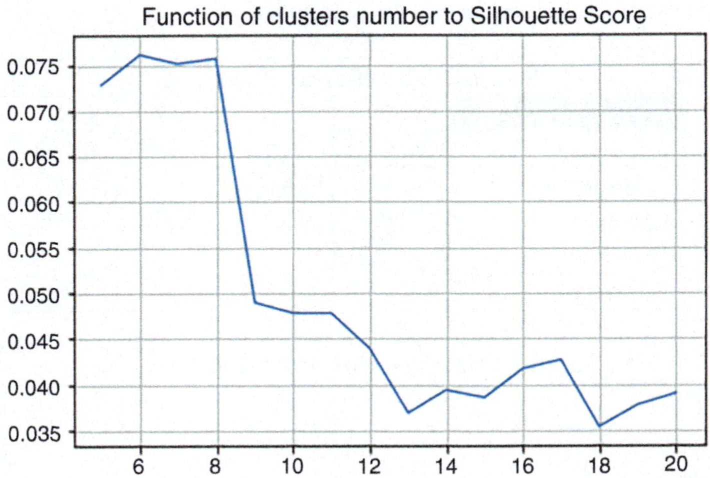

Fig. 6 Silhouette score graph

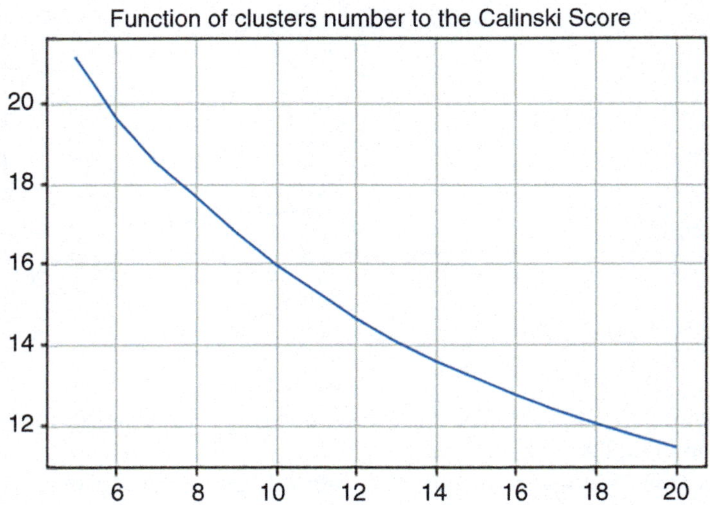

Fig. 7 Calinski Harabasz score graph

Collect User Tweets At this point in the paper we are ready to detect real user deceptive tweets but first, we need to collect a specific user tweets using Twitter API.

We were able to collect the tweets of the 44th president of the United State Barack Obama and the 46th Joe Biden. The tweets collected were posted in the 10th, 11th, and 12th of the year 2020. A total of 679 unique tweets were collected (120 tweets posted on Barack Obama's Twitter account and 577 tweets posted on Joe Biden's). The data collected with each tweet are:

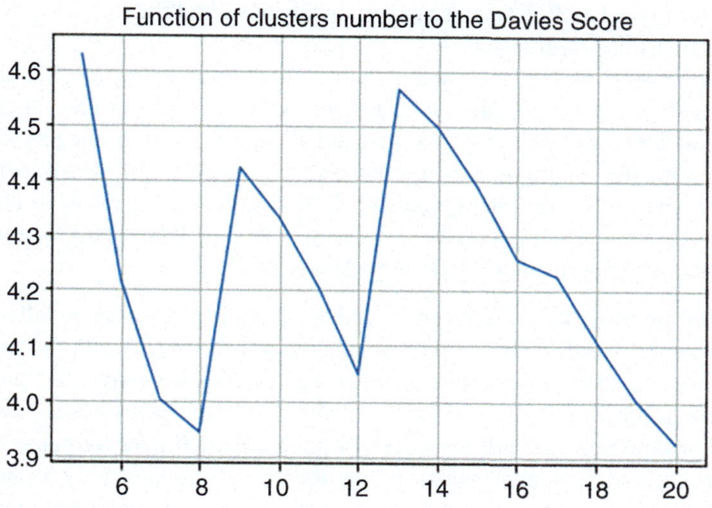

Fig. 8 Davies Bouldin score graph

Table 4 The resulted keywords of topics

Cluster	Keywords
0	(Vote, state, republican, polici, nation, Obama, people, candid, lawsuit, law)
1	(Privat, stolen, invest, illeg, tax, commerc, record, percent, income, job, budget, spend, commut, welfare, poor)
2	(Poverty, depart, analysi, economist, bankro, wage, deal, oversea, American, electr, firefight, privatesector, econom, director)
3	(Dodg, aid, bandaid, wealthi, crimin, benefit, agriculture, fraud, credit, tax-cut, income, treati)
4	(Regist, hike, exempt, texa, judg, childhood, cavity, obes, sexual, growth, abort, pump, import, investing, Fraud, trade)
5	(Retir, busi, trump, equal, rule, stock, nato, nomin, negoti, abort, break, nation, propos, nuke, prevent, punish, benefit, market, convers, putin, deleg)
6	(Terrorist, bush, public, subsidi, middleincom, deficit, decad, announc, sequest, putin, percent, rate, allow, labor, unveil, govern, secret, polici, clock, waiver, cut
7	(Pay, leav, inform, origin, cancer, oppos, American, care, , health, senat, percent, people, muslim, expand, hatch, propos, strip, requir, profess, child, advisori, nick, law, report, public, cosponsor)

- Text
- Date
- URL
- Username

Metadata (is_repley, replying_to, repley_count, retweet_count, fav_count, tweet_photos, has_video, has_qouted_tweet, quoted_tweet_url, has_external_url, external_ur).

Classifying New Tweets Using the gathered keywords and the collected tweets, we start to classify new tweets to the topics that we already defined.

- *Importing the data and preprocessing*: we proceed by importing the tweets as data frames and importing the list of keywords that identify the seven topics as a 2D array and delete any duplicates found. The same exact process of preprocessing the LIAR dataset is applied on Obama's and Biden's collected tweets to ensure a consistent result.

- *Classify tweets by keywords and synonyms:* using the keywords, we were able to classify both Biden's and Obama's tweets into our predefined categories by counting how many keywords we were able to match for each category keywords and finding the maximum matches between the categories. If the following array is the result of the matching process [2, 7, 5, 1, 3, 9, 0, 1] the result would be that the tweet belongs to the sixth topic (topic number 5 in Table 3) because we found the most matches using the sixth topic keywords.

To further improve the classification and to guarantee the best results we also matched the keywords' synonyms (provided by WordNet [22]) with the tweets and because finding a keyword should be better than finding a synonym, we proposed a point system for the matches found. Each keyword found will be equal to 5 points and each synonym found will be equal to 4 points. Furthermore we put a constraint on accepting a classification which is that the classification is only accepted if it has the maximum number of matches between all the topics and it has more than 8 points total. This constraint introduces a ninth category which is the tweets that we were not able to classify.

Comparing Obama's and Biden's dated tweets we were able to find a relation between both accounts. We notice that using synonyms helps decrease the percentage of unidentified topics by 29.5% in the case of Biden's tweets and by 28.6% in the case of Obama's tweets. We also notice that the dominant topic is topic_2 which talks about economy and politics (Table 4).

Noticing a close percentage between both results is a good indication for the accuracy of the topic classifier because as we will discuss in the next section the datasets (Biden's tweets and Obama's tweets) share a percentage of similarity between them as both Biden and Obama are republicans and were tweeting about similar topics in the period were the tweets were collected.

- *Deception Influence*: one of the factors that might affect the deceptiveness of a Twitter account is following or copying another deceptive account's tweets. By comparing Obama's and Biden's dated tweets we were able to find a relation between both accounts.

Using the dates of the tweets we compare all the tweets of Obama with all Biden's subsequent tweets and do the same for Biden's tweets with all the subsequent tweets of Obama. This experiment tells us who is following/copying the other. The result of this experiment showed that Biden follows Obama with 33.66% and Obama follows Biden with 22.22%. Performing this experiment again using the tweets synonyms showed that Biden follows Obama with 98.08% and Obama follows Biden with 91.76% which again proved that Biden is following Obama's lead in some topics (Figs. 9 and 10).

Training Classifiers At this step, we start to train a machine learning model on the different topics that we gathered and clustered. The model will detect deceptive

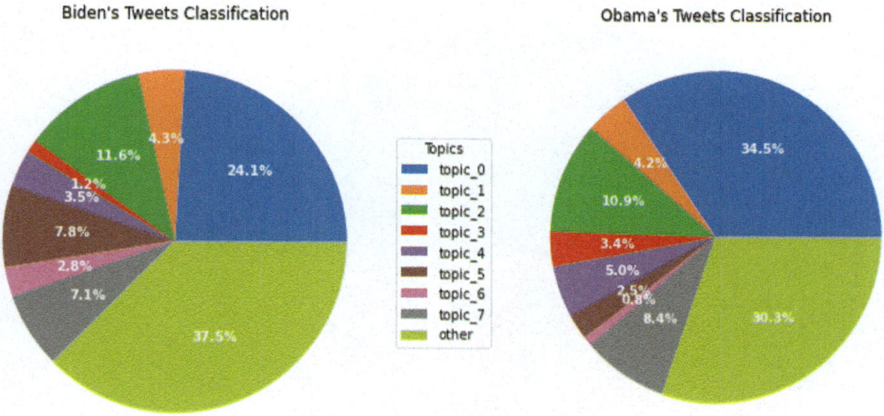

Fig. 9 Classification result of Biden's and Obama's tweets without using keyword synonyms

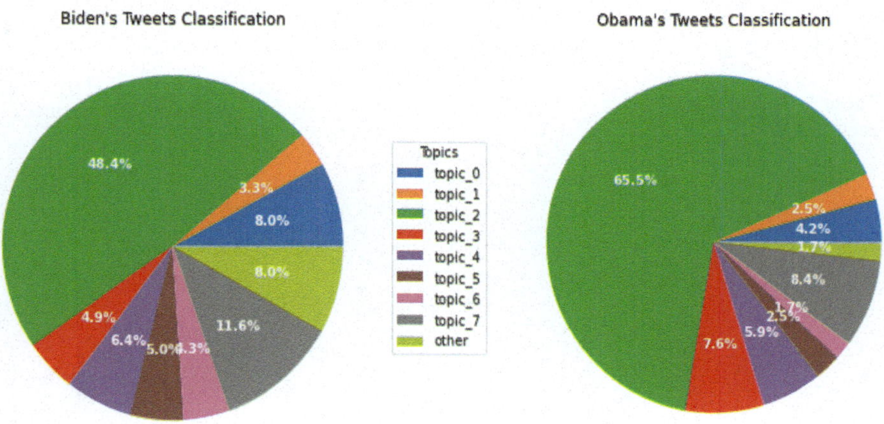

Fig. 10 Classification result of Biden's and Obama's tweets using keyword synonyms

tweets based on the writing style of each tweet and gather features that will define the deceptive style for each topic.

For this step, we used Logistic Regression classifier provided by scikit-learn [21]. We trained one model to classify all the news in the database before dividing them into eight topics first to be able to compare the difference between classifying the divided topics and not dividing them and measure how our process improved the deception detection.

One classifier was trained on 75% of the news in the LIAR database and was tested on the remaining 25%. The model scored an accuracy of 61% without dividing the topics and building multiple classifiers. After that, we trained eight different classifiers each on their respective topic from the previous steps, (Fig. 11) displays the distribution of each topic.

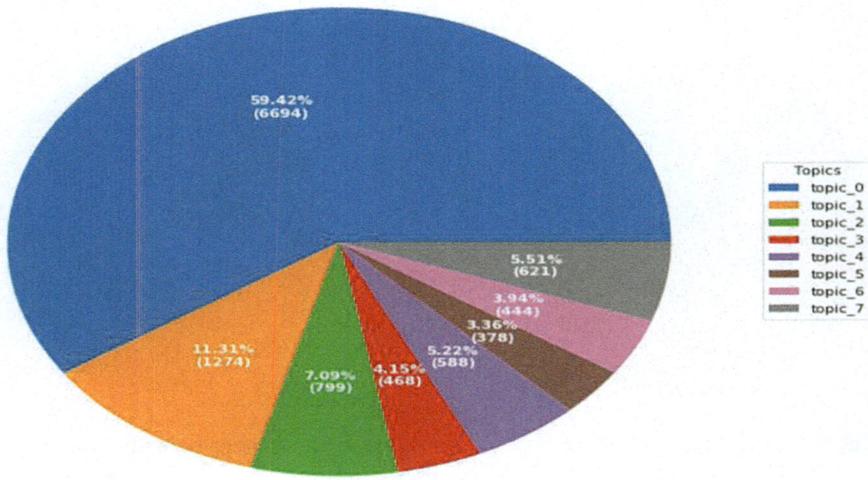

Fig. 11 LIAR database topic distribution clustered with agglomerative algorithm

Table 5 The accuracy of each classifier

Topic	Accuracy
0	60
1	69
2	78
3	54
4	74
5	60
6	61
7	63
Average	**65**

By training and testing eight different classifiers on these eight different topics we were able to obtain average accuracy of 65% which is a 4% increase in accuracy compared to training only one classifier, (Table 5) displays the result of each classifier for each topic and the accuracy average.

Calculating Credibility Score using the trained model and the tweet classifier we classified each user tweet and then detected whether its writing style is deceptive or not using the eight trained classifiers.

We calculated the credibility score of Obama's, Trump's, and Biden's gathered tweets during the 2020 US presidential elections. Figure 12 displays the classified topics of the three candidates.

We obtained a 23% credibility score, adding the error rate to the credibility score which is 35%. The final credibility score for Obama's Twitter account is 58%. Applying the same method to calculate the credibility score for Biden's Twitter account we obtained 59%.

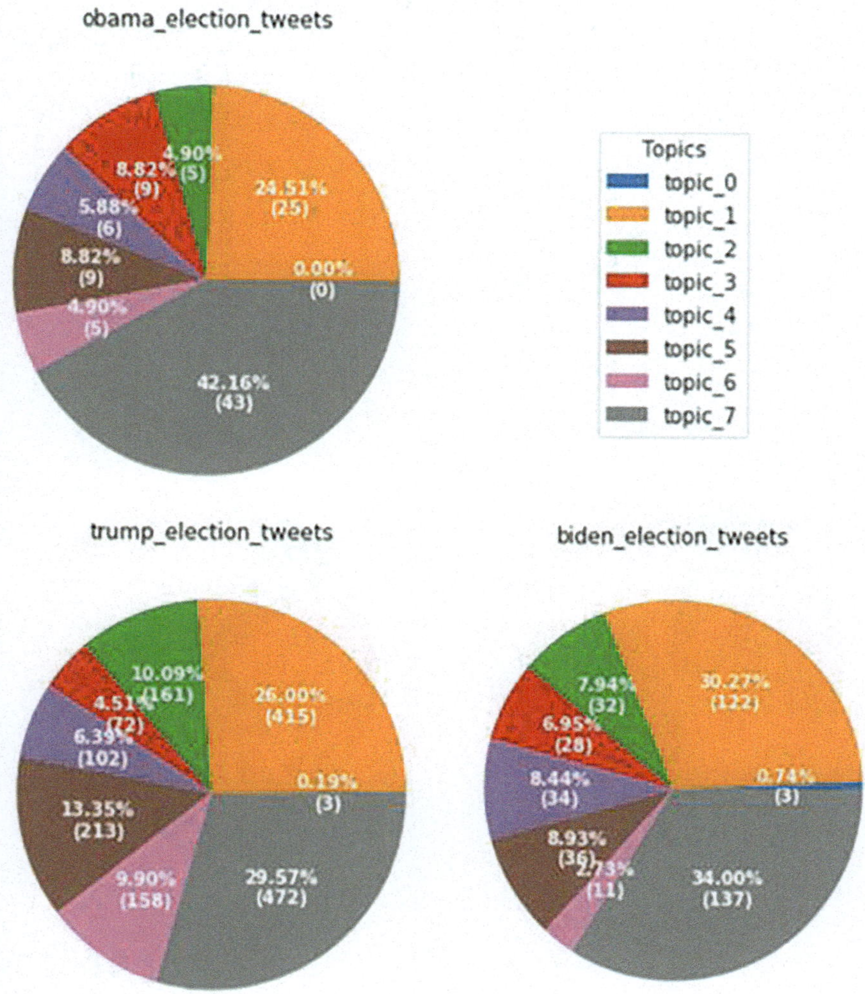

Fig. 12 2020 US presidential elections credibility scores for Obama's, Biden's, and Trump's Twitter accounts

This method can be applied to any Twitter account that has a sufficient number of tweets, we applied the same methodology on the Twitter account of the 45th president of the united states Donald Trump by collecting the account tweets [23] in the same period [23] of the other two presidents' accounts and the account obtained a 55.07% credibility score. Table 6 displays the credibility score for each account and the scores for each topic.

We continue by trying the same analysis on the 2016 US presidential elections in which Hillary Clinton and Donald Trump compete and Trump won, we collected all the tweets posted by Hillary's, Trump's, and Obama's Twitter account between

Table 6 2020 US presidential elections credibility scores for Obama's, Biden's, and Trump's Twitter accounts

Topic	Score		
	Obama	Biden	Trump
0	0.29	0.32	0.2
1	0.29	0.42	0.42
2	0.05	0.01	0.2
3	0.28	0.3	0.2
4	0.18	0.16	0.16
5	0.09	0.09	0.04
6	0.48	0.49	0.43
7	0.17	0.13	0.13
Total	**57.92**	**58.88**	**55.07**

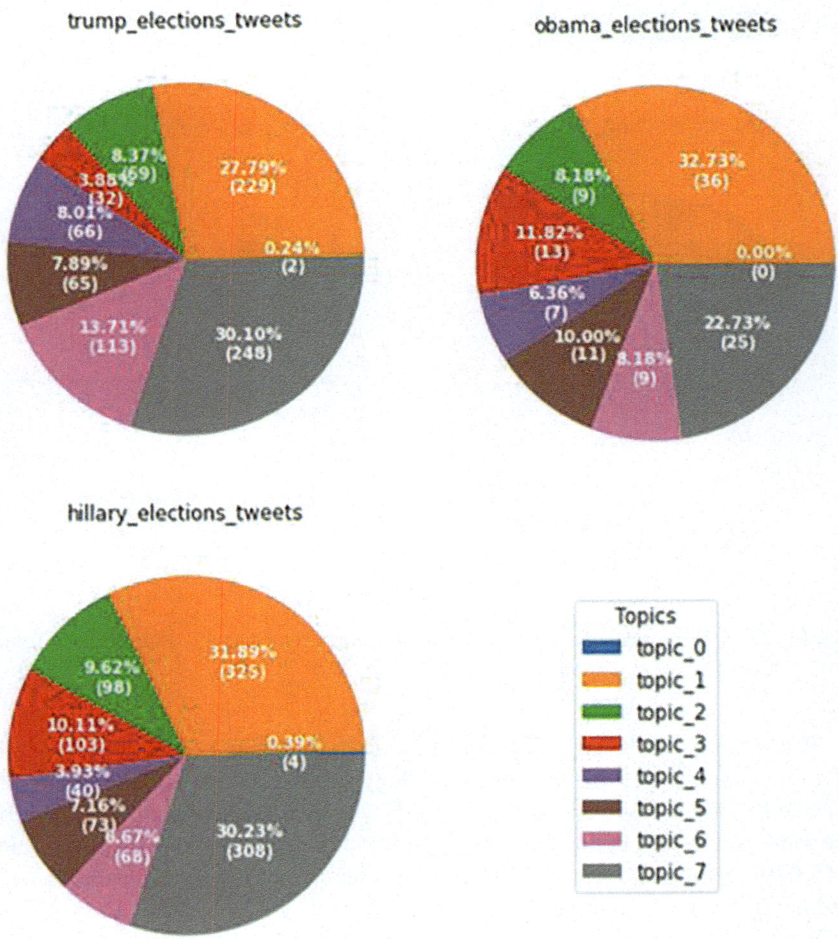

Fig. 13 2016 US presidential elections credibility scores for Obama's, Hillary's, and Trump's Twitter accounts

Table 7 2016 US presidential elections credibility scores for Obama's, Hillary's, and Trump's Twitter accounts

Topic	Score		
	Obama	Hillary	Trump
0	0.35	0.23	0.15
1	0.14	0.43	0.37
2	0.03	0.03	0.02
3	0.35	0.18	0.16
4	0.05	0.16	0.15
5	0.09	0.06	0.02
6	0.43	0.5	0.4
7	0.15	0.08	0.09
Total	**54.66**	**55.69**	**52.04**

2016-09-01 and 2016-12-30 which counted for 1608 tweets by Hillary, 174 by Obama and 1527 by Trump. Figure 13 displays the classified topics of the three candidates.

Using the same methodology and the trained model we classified their tweets and calculated the credibility results (Table 7) displays the credibility score for each account and the scores for each topic.

After that, we decided to further analyze Donald Trump Twitter account by collecting the account's tweets in different periods and calculating the credibility scores of these periods to see how the writing style changes during the account's lifetime.

We choose the following four periods to analyze:

1. Before presidency
2. During the first elections
3. During presidency
4. During second elections

The classification of topics during the four different periods is displayed in (Fig. 14) which changed from the first two periods to the last two where we can notice a decrease in topic_1 and an increase in the rest of the topics during the account lifetime.

The result we obtained shows that the Twitter account started with the worst credibility score and with time the credibility score increased which indicates that the writing style of the Twitter account increased in credibility with time. Table 8 displays the credibility score results of each topic and the total credibility score for each period.

4 Results

Performing the clustering step using agglomerative as described in the methodology section resulted in eight different clusters with a clustering accuracy of 66.798%.

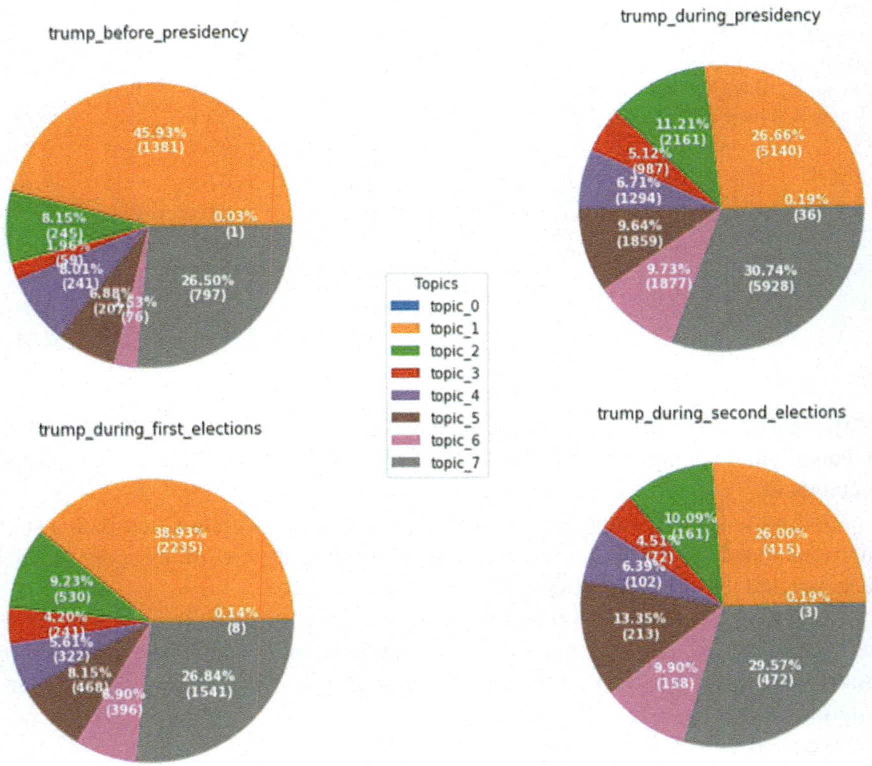

Fig. 14 Donald Trump tweets classification of four different periods

Table 8 Donald Trump credibility scores over four period

Topic	Score			
	Before presidency	First election	During presidency	Second election
0	0.28	0.2	0.2	0.2
1	0.34	0.36	0.33	0.42
2	0.02	0.03	0.03	0.02
3	0.12	0.16	0.2	0.2
4	0.09	0.14	0.14	0.16
5	0.03	0.03	0.05	0.04
6	0.43	0.45	0.4	0.43
7	0.07	0.09	0.11	0.13
Total	**52.31%**	**53.12%**	**53.18%**	**55.07%**

Table 9 displays different scores and metrics used to evaluate the agglomerative clustering algorithm when clustering for eight topics.

Table 9 Agglomerative algorithm clustering results of eight different clusters

Agglomerative algorithm clustering results when K = 8	
Accuracy	66.798
Silhouette score	0.076
Calinski Harabasz score	17.680
Davies Bouldin score	3.943

Using the extracted news clusters, we were able to obtain keywords that identify topics in each cluster by applying topic modeling and LDA algorithm. Table 3 displays the top keywords extracted from each cluster.

We gathered Obama's, Trump's, and Biden's tweets to test our methodology on them and classified each tweet to one of the eight predefined topics using the extracted keywords and keyword synonyms.

By training eight different classifiers (one for each predefined topic) on the LIAR dataset we were able to improve the deception detection accuracy from 61% to 65% and we were able to calculate a credibility score for Obama's, Biden's, and Trump's Twitter accounts during the 2020 US elections which resulted in 58%, 59%, and 55% scores respectively (Table 6) and calculate a credibility score for Obama's, Hillary's, and Trump's Twitter accounts during the 2016 US elections which resulted in 55%, 56%, and 52% scores respectively (Table 7).

We also analyzed different periods of President Donald Trump's Twitter account which resulted in 52% in the first period, 53% in the second period, 53% in the third period, and 55% in the fourth (Table 8) which also proved that the credibility of the account's writing style is improving over time.

5 Discussion

Although the accuracy of the clustering step was low, we obtained well-defined keywords for each cluster when using topic modeling to extract the cluster keywords. We also believe that we can enhance this accuracy in the future by tweaking the parameters or searching for a better clustering algorithm.

Our method improved the style based deception detection of fake news from 61% to 65% based on the assumption that fake news deception style changes from one topic to another, although the improvement percentage is only 4% the methodology we used has shown that our assumption is correct and that this percentage can be further improved in the future by using other flags and information about the Twitter account like considering the network that the account is in, the tweets propagation behavior and the interaction of other users with the specified Twitter account.

13. R.J. Fisher, Social desirability bias and the validity of indirect questioning. J. Consum. Res. **20**(2), 303–315 (1993)
14. A. Ward, L. Ross, E. Reed, E. Turiel, T. Brown, Naive realism in everyday life: Implications for social conflict and misunderstanding. Values Knowledge **1997**, 103–135 (1997)
15. C. Silverman, This analysis shows how viral fake election news stories outperformed real news on Facebook. BuzzFeed News **16**(2016) (2016)
16. S. Vosoughi, D. Roy, S. Aral, The spread of true and false news online. Science **359**(6380), 1146–1151 (2018)
17. Zhou, B., Pei, J.: OSD: An online web spam detection system. In: Proceedings of the 15th ACM SIGKDD international conference on knowledge discovery and data mining, KDD, vol. 9 (2009)
18. X. Zhou, R. Zafarani, Network-based fake news detection: A pattern-driven approach. ACM SIGKDD Expl Newsl **21**(2), 48–60 (2019)
19. Shu, K., Mahudeswaran, D., Wang, S., Lee, D., Liu, H.: FakeNewsNet: A data repository with news content, social context and dynamic information for studying fake news on social media. arXiv preprint arXiv:1809.01286 (2018)
20. Sitaula, N., Mohan, C.K., Grygiel, J., Zhou, X., Zafarani, R.: Credibility-based fake news detection. In: Disinformation, Misinformation and Fake News in Social Media: Emerging Research Challenges and Opportunities. Springer (2020)
21. S. Bird, E. Loper, E. Klein, *Natural Language Processing with Python* (O'Reilly Media Inc., 2009)
22. About WordNet.: Princeton University (2010)
23. Brown, B.: Trump Twitter Archive. www.thetrumparchive.com (2016)

Detecting Trending Topics and Influence on Social Media

Miray Onaran, Ceren Yılmaz, Feridun Cemre Gülten, Tansel Özyer, and Reda Alhajj

Abstract In today's world, social media platforms have a significant potential impact on individuals in various areas and issues. Twitter is one of these social media platforms with its wide communication network. In addition, Twitter has become a popular social media platform for sharing and communication in the field of health. It has the potential to have a significant impact on individuals in the field of health, as it is a platform used by both the public to obtain information and the scientists to share information. At this point, the analysis of the tweets and tweet owner on Twitter in the field of health for Turkey has been considered as an important topic to be focused on. Accordingly, the main purpose of this study is identifying and predicting trending healthcare topics and influential people in Turkey during a specific period. By doing this, important health issues and problems in Turkey will be determined. Accounts with potential impact in the field of health will identify identified. In line with this goal, an efficient model as a method is proposed for determining healthcare trending topics and influential people. The proposed method to cope with this problem includes two aspects, namely utilizing the n-grams algorithm to identify trending topics and employing social network analysis to identify influencers. In addition, the developed framework has an interface design to display trending topics and influential people for the desired time interval.

Keywords Trending topic · Influential people · Twitter · Social media · Healthcare

---

M. Onaran · C. Yılmaz · F. C. Gülten
Department of Computer Engineering, Istanbul Medipol University, Istanbul, Turkey

T. Özyer
Department of Computer Engineering, Ankara Medipol University, Ankara, Turkey

R. Alhajj (✉)
Department of Computer Engineering, Istanbul Medipol University, Istanbul, Turkey

Department of Computer Science, University of Calgary, Calgary, AB, Canada

Department of Heath Informatics, University of Southern Denmark, Odense, Denmark
e-mail: ralhajj@medipol.edu.tr

S. T. Özyer, B. Kaya (eds.), *Cyber Security and Social Media Applications*,
Lecture Notes in Social Networks, https://doi.org/10.1007/978-3-031-33065-0_8

1 Introduction

The increase in the use of computers and smartphones together with the wide access to the Internet has caused social media sites like Twitter, Facebook, and Instagram to have a very large place in daily life around the world. Social media platforms have been able to affect many areas of our lives. In addition, social media platforms have become important sources in the field of knowledge discovery for informative decision making. These platforms cause significant effects on people in terms of obtaining information, receiving news, sharing ideas and opinions. It is ensured that individuals from various countries, age groups, and occupational groups interact with each other by communicating with each other. Social media platforms are social interaction areas where people generate, share, and exchange ideas on a wide variety of topics. Social media platforms are global where distances are broken, and people can relate to societies. All this communication has a significant impact on people. These platforms can be described as important tools that have an impact in areas such as interpersonal interaction, social behavior of individuals, and social issues. Twitter is one of the most widely used social media platforms. Twitter is an information network with a large user base; it is based on the sharing of short messages called tweets limited to 280 characters [1, 2]. People can share information and communicate via Twitter on a wide variety of topics such as agenda, history, religion, politics, environment, etc. For instance, Twitter is a social media platform widely used for sharing and communicating health-related information.

Twitter can be described as a data source of increased popularity. Accordingly, it can be qualified as a versatile platform for research in the healthcare domain. Scientists and healthcare professionals commonly prefer Twitter to disseminate and share scientific information. Also, Twitter has become an accepted resource for the public on health-related issues due to its easy access. In fact, Twitter is the most used social media platform professionally by scientists [2]. Twitter has the potential to influence large audiences in the field of health. Influential accounts own the potency to influence health issues such as raising awareness about health problems, advocating health, and causing health-related change at the social level. It has the potency to feed into the development of public health subjects at individual, societal and social levels. In addition, influential people have significant potential in health issues and interventions, especially in promotion and development [3].

Considering the widespread use and impact potential of social media, as well as Twitter's potential in the field of health, it was deemed necessary to address this area for Turkey as well. Identifying and predicting trending topics in the field of health in Turkey is important in terms of detecting problems in the field of health. In addition, identifying and predicting influential accounts in the field of health is important in terms of mass and social interventions and referrals. Considering all this, the main purpose of this study is identifying and predicting trending healthcare topics and influential people in Turkey during a specific time period.

The tasks accomplished by this study can be summarized as follows:

- Identifying and predicting healthcare trending topics in Turkey during a specific time period, thus identifying health problems on a country basis or regionally.
- Identifying influential accounts (persons or organizations) in the field of health in Turkey for a specific time period, thus revealing their potential owners in the fields of health-related incentives, development, and intervention.
- Proposing a model for determining healthcare trending topics and influential people.
- By developing an interface design, a certain time period can be selected, and trending topics and influential people for this time period can be displayed.

The targets of this study can be summarized as follows:

- To draw attention to the impact potential of Twitter in the field of health and to raise awareness.
- Analyzing the status of health problems in Turkey. To draw attention to the solution of health problems that need to be improved and prioritized.
- Encouraging future studies involving healthcare trending topics in Turkey in Twitter and to be able to guide relevant future studies.

When the literature was examined, it was observed that there are several methods and studies for detecting trending topics and influencers on social media. However, the features they considered for the data are limited. Thus, efficient methods will be proposed in the project by increasing the number of features while investigating the tweets of users. Furthermore, it was discovered that influencers were identified by ranking them based on their number of followers and mentions. In this study, more features have been used for scoring and detecting the influencer users such as retweet, reply, and favorite. Additionally, to the best our knowledge, considering the literature in such a study about detecting influential people and a topic on the health subject has not been conducted in Turkey.

The rest of this chapter is organized as follows. Section 2 is related work. Section 3 described the methodology. Section 4 covers determining trending topics and influential people from open source data. Experimental results using the data collected from tweeter are included in Sect. 5. Section 6 covers the discussion. Section 7 is conclusions and future work.

2 Background and Related Work

2.1 Background Information

Social Media This covers the visual and auditory tools that convey all kinds of information to individuals and society and have three basic responsibilities such as entertainment, information, and education. Social Media is an online networking platform where users create their accounts and share their own opinion or media. Some of the popular platforms are Facebook, Instagram, YouTube, Pinterest, Twitter, etc. It has positive effects such as recognizing and learning different ideas, quick access to information, acquiring new business contacts, promoting, and marketing the products for the companies, etc. On the other hand, it has some negative aspects such as it may create addiction, personal information may be collected by unknown people, also it may affect the spread of misinformation.

Influencer People who can reach, influence, and direct the thoughts of other people on social media are called influencers. They have an impact on marketing strategies. They are also effective in the emergence of topics that will create trends in society. Influential users and trending topics can be related to any topic such as technology, fashion, economy, health, etc. For example, the ideas shared by the influential users in the vaccine campaigns developed for the covid-19 virus may have changed the viewpoints of the peers against the vaccine. Therefore, the effects of these influencers and the trending topics on social media can be the subject of research to study how people react to and ride the wave of the trend, how the influence may diminish, etc.

Twitter: Twitter has become a popular social media platform on which accounts can share their ideas, media, etc. Users can post tweets which are limited to 280 characters. A tweet may be posted to other accounts by including "@" and hashtags (a hashtag is a keyword, or a phrase used to describe a topic or a theme by placing the symbol "#" in front of them) [4] (https://help.twitter.com/tr/using-twitter). Retweet and favorite are features that create an effective interaction between users of the platform. As the number of followers of an account increases, its popularity also increases. However, owner of an account with many followers may not be an influential person. Thus, it is possible to investigate retweet, mention, favorite and key words of the tweets by considering the relationship of the users to detect influential users and trending topics.

Twitter API Twitter permits authorized researchers to get public tweets and accounts. Twitter API is a tool which provides getting data from unprotected accounts or users. The app coded to post queries using Twitter API can share one of the quotes from a pre-constructed list for a user each day, then search for tweets about a user and save them in a file. Twitter data differs from data shared by many other social platforms in that it reflects information that users choose to share publicly. Twitter API supplies reaching public data that users share with other people on Twitter (https://help.twitter.com/tr/rules-and-policies/twitter-api).

Graph It is a network structure which consists of nodes connected by edges to reflect the relationship between entities pertaining to the investigated domain. Links in a graph may be directed from a source to a destination leading to directed graphs. Twitter is an example of a directed graph. Alternatively, graphs with undirected links are common in several applications, Facebook is an example of an undirected graph [5].

Flask It is a web framework from Python programming language. It has a collection of libraries and codes that can be used to construct websites. With these features, there is no need to do everything from scratch [6].

2.2 Related Work

There are various studies about detection and identifying trending topics and influencers. These related works are summarized and analyzed below.

Fiaidhi et al. [7] discussed some issues related to Twitter trending topic. They focused on the problem, i.e., although Twitter offers its users with a list of the most trending topics, generally it is hard for understanding the content of trending topics when it is far from personalization. Based on this, a twitter client has been developed to filter personalized tweets and trending topics. RSS (Really Simple Syndication) feeds allow for personal preferences to be included. The developed twitter client allows users to group tweets with captions that are explicitly or implicitly defined based on their preferences. Thus, Twitter users will be able to view the elements they are interested in more easily. The Levenshtein Distance algorithm and the LDA (Latent Dirichlet Allocation), were used to perform topic clustering. United States (New York) and Canada (Toronto) were selected for the location status of the tweets. Geocoding API is used to receive data with this location information. Streaming data was collected using Twitter4j API. A total of 2,736,048 (economy 1,795,211, education 89,455, health 390,801, politics 60,265, sports 400,316) tweets were collected [7]. This study simplified the search and identified the target tweets by finding personalized trending topics and grouping tweets by consistently clustered trending topics for more straightened exploration. However, this study only focused on shaping tweets based on personal preferences.

Adegboyega [1] examined the effect of social media. This study was directed to examine the effects of social media and its social behavioral effects on students. It is emphasized as social media has many good as well as many unpleasant sides. Children can be badly influenced by social media when they are not well-watched or monitored. Students and young people can be affected by the negative effects of social media. This study aims to examine and analyze the effect of social media on students' social behavior with the observations of primary school teachers in Ilorin metropolis, Nigeria. Based on the stated problems, a research question was determined about the effects. The descriptive survey method was used in the study as being quantitative research. Surveys were used to collect data within the scope of this method. The purpose of using the surveys is to determine the effect of social media on the social attitude of students. Simple random sampling technique among all 60,054 primary school teachers in Ilorin Metropolis 200 teachers were selected and participated in the survey. Afterward, 3 null hypotheses were assumed and tested. According to the tests conducted, the hypotheses were not rejected, and according to the opinions of the teachers, there was no statistically considerable difference in the effect of social media on social behaviors according to gender, age, and education level [1]. In the study, the effect of social media on people and students in today's world was emphasized. Some statistical methods have been applied. On negative aspect of this study is that only statistical methods have been used. Apart from that, information in areas such as tweet review, data extraction from tweets, tweet analysis, which are related to our project, were not emphasized.

Albalawi and Sixsmith [3] mentioned the effects of social media platforms on public health. The main contribution of this study can be seen as the promotion and development of public health in Saudi Arabia via Twitter. The authors aimed to use ways of applying and comparing multiple impact indicators in the field of public health for the implementation of promotion and development efforts. In line with this, the most influential Twitter accounts have been specified, characteristics (corporate or individual) and classification of the accounts have been specified. To determine the most popular 100 Twitter users 4 different scoring tools, namely, primary list development, filtering, classification, and analysis were used. In the primary part, a list was obtained using Tweepar and SocialBakers sites that service a list of the top accounts for the country. The final list consists of 182 users. In the filtering stage, a filtering process was implemented for the primary list. The list is included in four several influence-scoring tools. Social Authority by MOZ, PeerIndex, Kred, and Klout was used to get an influence score of 4 for each Twitter account. In the classification part, the accounts were first divided into 2 groups, individual and organizational. Then, they were divided into 2 groups as males and females. Third, 10 groups were classified as religious, sport, media, political, twitter, health, new media, services, royal family, and unknown. The final list consists of 99 accounts. In the analysis part, simple calculations, ratios, and percentages were utilized to examine the data and identify the best influential accounts.

Yeung et al. [2] worked on Twitter for discoveries in the healthcare domain. They mainly tacked the following problems: (1) lack of bibliometric analysis of Twitter usage in terms of health-related research, (2) lack of identification of relevant scientific literature, and (3) lack of quantitative analysis. The main virtue of this study is that the scientific literature on the use of Twitter in the health context was researched and analyzed using a bibliometric approach. Since there were not many studies that performed bibliometric analysis on the use of Twitter in the field of health, Yeung et al. were targeted to determine scientific literature and make quantitative analysis by obtaining a new perspective. In addition, the study intended to assist researchers who will work in this field in identifying collaborative partners and journals where research results can be published. 2582 articles were retrieved from Web of Science and analyzed to find Tweets and articles related to health. Basic bibliographic information was acquired, and more detailed analysis was performed using VOSviewer, a special bibliometric software. Term map and keyword map synthesis were created. The purpose was to visualize repeated words in titles, abstracts, and keywords. The study draws attention to how social media platforms such as Twitter can be utilized for posts about health.

Winatmoko and Khodra [8] worked on automatic summarization. The definition of the problem can be summarized as follows: Twitter provides a list of trending topics to its users. Users cannot understand the meaning of the topic because the topics are presented only as a list. In the article, it is aimed to define the characteristics of the trending topic. Studies have been made on automatically explaining the trending topic based on tweet collections. The method developed by the authors automatically creates a representation for the related trending topic. The authors target to identify trending topics that required explanation. They also target the generation of a tweet summary to describe trending topics. The study focused on generating explanations for trending topic in Indonesia, where no similar research has been found. The trending topic explanation method suggested in the article is carried out in 4 steps: pre-process, topic categorization, sentence extraction, and sentence clustering. For the data, 300 random tweets were collected from Twitter on 1–2 January 2013. 8 trending topics were determined.

Yang and Peng [9] focused on Weibo, which is a social media platform widely and actively used in China. The tackled problem can be summarized as determining whether closing the trending topic section influences the post and its interactions. This study is related to digital gatekeepers and trending topics in social media. User interaction data of 36,239 posts over a 3-week period was analyzed, i.e., using a natural experiment on Weibo. Data was collected from 50 accounts for a period of 4 weeks. A python script was written using the selenium package for post collection. The data was analyzed based on 4 aspects, namely, presence of trending topics, news engagement, top news items, and popularity of news accounts.

The goal of the study described in [10] is determining influencers and their opinions about the products based on their Twitter posts. This analysis gives information about influencers; and their information can be used for e-commerce and promotions. Business companies are interested in this study because they can use the analysis results to increase the performance of the commercials [10]. The authors used multilayer network, tensor model and SocialAU algorithm. They investigated all sectors, not just marketing or cosmetics.

The methodology of this study consists of four main steps: collecting tweets related to the decided products set and relevant data related to influencers, building multiple-layer network and tensor model by using the collected data, identifying influencers by using SocialAU algorithm, and identifying dominant products and related perceptions. Their perspective is showing that the idea of the general methodology can be applied to different product types. The literature review shows how this methodology is used for other studies. However, the other studies focus on specific topics. So, this study's main approach gets a result with more topics [10].

In some other studies, a graph theory-based method is generally used. However, the method used derives the relationships by analyzing the information from tweets' content. In this method, the concept is not analyzing only network topology, also related to opinions expressed on topics of interest [10]. Twitter API is used for downloading and collecting tweets about targeted products. Each collected tweet is labeled with the mentioned products' name. Author name is also captured. Three layers network was built as a network topology. Each layer of the network includes one of these topics, users, products, and keywords. In the study, SocialAU algorithm was used to find the most authoritative users and their tweets regarding products. Influencers were sorted according to popularity related to a specific topic. User's score is used as a mark to determine the popularity of influencers. At the same time, keywords show the subject of the shared tweets [10]. Data selection is an important aspect for this type of study.

The study described in [11] is concentrated on collecting, ranking by the number of views, and tracking breaking news in Twitter. The examined single message and timeline aspects of identifying breaking news. There are two types of single message aspects, namely text-based and emotions. The authors investigated just text-based type single message aspects. The identification can be done by keywords which are nouns and verbs. Some specific nouns and verbs are selected in the study, including names of famous places, people, and events. As verbs, the keywords selected include fire, crash, bomb, win etc. Most important or mostly read tweets and their topic with hashtags are grouped and ranked. The information gathered from the collected tweets, is analyzed and transferred to the application. Data is collecting using Twitter API. #Breakingnews hashtag is used while data is collecting.

The work described in [12] focused on proposing a different approach called Weighted Correlated Influence (WCI) for identifying influencers on Twitter. Other studies used only a single parameter method while WCI is a multiple parameters method. This is a new perspective to the identification methods. The basic methodology of WCI is to calculate the influence score for each chosen user in a network. WCI is based on a graph model where nodes and edges represent users and their relationships based on tweets, retweets, mentions, replies, follow-ups, and followers. Twitter's REST API has been used for data collection. At the end, algorithms and matrix normalization were applied to the data according to WCI process. Public data was used in the project. It was collected separately using #CoronavirusPandemic and #DelhiViolence hashtags, leading to 15,018 and 18,473 number of users, respectively.

There are several factors to consider for topic detection and tracking. The work described in [13] focused on these factors and how they may affect the detection results. Six different detection methods were investigated, namely, Latent Dirichlet Allocation, Document- pivot topic detection, Graph based feature-pivot topic detection, Frequent pattern mining, Soft frequent pattern mining and BN-grams. The collected data was analyzed with all these methods and the results were compared with each other. Three different data topics were selected: Duper Tuesday, FA Cup and Election. Each data set was collected based on a different time slot. For example, Duper Tuesday data was collected on one hour range, FA Cup data was collected based on one minute range, and Election data collected was based on 10-minutes range. The reason for the different time slots per data set is due to the nature of the application from which the data was captures. For instance, football (soccer) matches have a different nature than the election. LDA was identified as the best method for the noisy data sets.

The work described in [14] introduced different methods to automatically detect outstanding events. The main analyze of this method is Story Detection. As the traditional approach of FSD gives poor performance and high variance with locality sensitive hashing, they adjusted the FSD method with the LSH process. Traditional FSD represents documents as vectors. Each new document is compared with previous documents and the similarity between them is analyzed to classify documents accordingly. The modified version of FSD virtually eliminated variance and significantly improved the performance of the system. In the modified FSD, the same settings of the UMass system were used. Some steps of this method were to limit the system to preserve only the top 300 features and set two LSH parameters. The number of hyperplanes k provided a tradeoff between the time spent computing the hash function. The results are compared with manual labeling results. However, the collected data was nearly 160 million posts. So, some part of the data was manually labeled and compared with the study result. One of the details of the study is that # and @ symbols were not included to the research. The reason is that researchers did not want to have the result affected by some platform specialties. This method can be used on different platforms. The data was collected for 6 months. The results of this methods were ranked with different threads, including baseline, number of users and entropy with users.

In the study described in [15], researchers tried to determine new criteria to identify social media influencers. They introduced three different features to use in analyzing and determining influencers. These features are the number of followers, social authority, and political hashtags. In this study, researchers focused on Saudi influencers. These selected features matched with three primary metrics. The top 10 political twitter users were found by the three metrics by the filtering process. This study concentrates on three stages, namely gathering data, filtering, and semantic analysis. While collecting the data, the users were selected according to the primary metrics. In the second filtering process, the collected users were filtered with different more personalized criteria, including account needs to be personal, not business or another type of account, the account holder should be Saudi, and it should be an actively used account. While analyzing the filtered accounts, political hashtags were used. At the end of the study, the top 10 Saudi twitter users were identified and analyzed with these steps. In this study, a public source data was used. The data was not private. As a result of this study, it was realized that selecting features while filtering can affect the results.

In the work described in [16], the authors proposed a method to detect influential users. It was stated that influential users are generally detected by considering the retweet, followers, or measurement of centrality. The study highlighted the fact that gathering communication relationships among users, users' profile features and link analysis approach are used to find influential users. The compared methods are mentioned as UIRank, FansRank, ToRank, and Retweet Influence. In the compared methods, few features were considered while detecting influential users in a trending topic. For example, number of followers was considered in the FansRank method. Also, the count of retweets of a given tweet were considered while detecting influential users. Precision, recall and F1 score were used to compare the proposed method with others. The result of each method was reported as graphs. It was observed that for the Fans Rank method recall, F1 score, and precision were at the lowest value. Also, Retweet Influence had low values of precision than the proposed method. As a result, it may be concluded that number of followers and tweet counts alone do not indicate whether a given person is an influencer. It may be claimed that to determine an influential user using the proposed method [16], more relationships and traits should be used. The authors scored influential users by considering not only the follower and number of tweets, but also the communication relation.

The work described in [4], the authors pointed out how influential people have a lot of impact on social media users. The effects may change marketing strategies of a company or an industry. Thus, the authors focused on detecting influential users related to fashion technology using centrality measures, such as eigen and degree centrality. After performing social network analysis, influential users were scored. Centrality was used for understanding the network. Graph theory techniques were used to compute the essentiality of a node in the network. By considering the number of followers and friends, users and their relationships were analyzed. 1000 topmost tweets that included the fashion technology mentions were collected and 80 users were selected by considering the retweet information. It was determined that, based on the information acquired, 90% of the accounts were influenced by fashion technology. The authors mostly focused on the text of the tweets. The amount of mentions and retweets was evaluated. To reduce the interaction, the authors used API call load data to retrieve all tweet IDs of the given list of 100 tweets based on the fashion industry and stored them in a variable called (data frame) [4]. The dataset used has a good number of features, but the authors mostly focused on tweet_id and tweet_text. Thus, the number of focused features can be increased. The user with the highest score is considered as a better influencer in the network about a particular topic or field. In our work described in this chapter, influential people and trending topics can be scored by considering some features as input.

Demir and Ayhan [17] examined the relationship between the policy agenda and Twitter's public agenda using the social network analysis method. Agenda setting theory and social media issues were mentioned, and a comparison of traditional agenda setting model and network agenda setting model was performed. #ayasofyacamii tag posted by Twitter users was examined. Regarding the decision to make Hagia Sophia a mosque again, a hashtag was created on Twitter and approximately 800 thousand tweets were sent to the hashtag #ayasofyacamii between 10 July and 21 July 2020. After analyzing the data of 12 and 13 July, the first day of the discussion on Twitter, the tweets sent at intervals on 17 July and 19 July were examined to determine whether different agenda setters were effective on the public agenda. In addition, on July 24, 2020, the hashtag came to the fore again and approximately 500 thousand tweets were sent. The data was collected using NodeXL software, which functions as an add-on to the Microsoft Excel program, and has an interface that helps to collect the tags or words used by users on Twitter. In the findings and interpretation part, the general view of the networks formed regarding the #ayasofyacamii hashtag, the findings including the relationship between the public agenda and the policy agenda, the analysis results and interpretation of the actors that dominated the network and the agenda setters were presented. To determine dominant actors of the agenda, i.e., to determine how powerful the actors in the network are on the information flow, they considered the values of the betweenness centrality. The eigenvector centrality method was used to evaluate agenda setters.

Bakan [18] examined corporate Twitter accounts of the forty best art schools in the world between 2018 and 2019 according to the social network analysis method. Internet and social media, Twitter, social network analysis, centrality measures, sample social network analysis studies were mentioned. Twitter accounts were analyzed according to social network analysis; tweets, hashtags, number of followers, followed, shared photos and information in the videos were taken into account in the analysis. Frequency and mean statistical processes were applied in the analysis of the data. Sentiment analysis, which is one of the content analysis methods, was used to examine the views on the agenda in art schools as positive, negative or neutral from a class perspective. For the sample of the study, 40 schools determined according to academic criteria were taken into consideration. Information such as content, followers, likes and comments in the corporate Twitter accounts of these schools were systematically collected for a 1-year period (2018–2019). NodeXL program was used for numerical analysis and visualization of the network structure. The results of measuring density, degree of centrality, betweenness and closeness centrality, and the roles of actors in the network were analyzed. The distribution of interaction networks on twitter, centrality measurements of universities' twitter networks, intuition analysis results of universities' twitter contents were presented and explained.

3 Methodology

The work described in this chapter has two main purposes, namely, to identify trending topics and influencers. To achieve these two main objectives, n-grams and social network analysis have been used, respectively. The n-grams algorithm was chosen as the method to determine trending topics based on tweets. Social network analysis was utilized to identify influential people based on tweeters.

3.1 The N-grams Algorithm

N-grams is a machine learning process mainly used for text processing. N-grams are basically continuous sequences of symbols from the alphabet considered in the analysis. For the case described in this chapter, the alphabet consists of words, letters, or symbols. The operation of the n-gram algorithm is to separate word groups in the given input according to the type of the n-gram used. Word groups are identified and stored based on the value of n such that each consecutive list of n words form one group; this may be based on unigram (n=1), bigram (n=2), trigram (n=3), etc. to illustrate the process, consider the following sentence "The Omicron variant spreads more easily than the original virus that causes #COVID19." (https://web.stanford.edu/~jurafsky/slp3/3.pdf).

For unigram case (n=1), the sentence is separated and stored as:
"The", "Omicron"," variant", "spreads", "more", "easily", "than" "the", "original", "virus", "that", "causes", "COVID19". For bigram case (n=2), the sentence is separated and stored as:
"The Omicron", "Omicron variant", "variant spreads", "spreads more", "more easily", "easily than", "than the", "the original", "original virus", "virus that", "that causes", "causes COVID19". For trigram case (n=3), the sentence separated and stored as: "The Omicron variant", "variant spreads more", "spreads more easily", "more easily than", "easily than the", "than the original", "the original virus", "original virus that", "virus that causes", "that causes COVID19".

Punctuations are removed from the sentences before the n-gram algorithm is applied. Accordingly, the hashtag symbol '#' has been excluded from further consideration in the example. We decided to use n-grams in our work after we realized the successful usage of this technique in a variety of domain like the one covered in this chapter, including, spam filtering, auto completion of sentences, auto spell checking, etc. For autocompletion of sentences, a model is build based on a training set of n-grams extracted from some complete sentences, then it predicts and suggests some words for incomplete sentences. Deciding on the value of n in the process may be driven by the application domain. For instance, it is more effective to use 30-gram and 4-gram for spam filtering (https://www.analyticsvidhya.com/blog/2021/09/what-are-n-grams-and-how-to-implement-them-in-python/).

In this chapter, the n-gram algorithm is used to determine trending topics. The data analysis with the n-gram algorithm proceeds as follows. After separating the word groups, they are stored and the words in each group are counted. Then, word groups are sorted in descending by the number of words leading to trending topics. To get more meaningful and effective results, in this chapter stop words were filtered suing the special python library "stop_words ()". Since the purpose of the chapter is determining trend topics, it was realized that unigram and bigram gave more proper.

3.2 *Sentiment Analysis*

Sentiment analysis involves identifying and analyzing a text to determine if it is positive, negative, or neutral as reflected in the example shown in Fig. 1. It is used in Natural Language Processing for different purposes. Companies use sentiment analysis for customer feedback. One of the popular applications of sentiment

Fig. 1 Example of the sentiment analysis (https://www.datacamp.com/community/tutorials/simplifying-sentiment-analysis-python)

	news	sentiment
0	Reporting in accordance with the merged busine...	neutral
1	It is the last smartphone running Maemo 5 , wh...	neutral
2	A meeting of Glisten shareholders to vote on t...	neutral
3	Ruukki Romania , the local arm of Finnish meta...	positive
4	stores 16 March 2010 - Finnish stationery and ...	negative

analysis is social media where the posts can be detected and filtering as positive, negative, and neutral (https://realpython.com/python-nltk-sentiment-analysis/).

Sentiment analysis follows some steps to get proper results. First, the input data is specified; then stop words and some punctuations are filtering. In the next steps, negation handling, stemming and classifications are processed. However, for this analysis python has a library and a specific function which can be used when people want to apply sentiment analysis to their own data. Classification may be applied for cases where there exist extra preferences. Classifiers such as Naive Bayes, k-NN, and decision tree can be used for sentiment analysis (https://www.datacamp.com/community/tutorials/simplifying-sentiment-analysis-python).

In this chapter, the NLTK library has been used to process the sentiment analysis. The NLTK library contains various utilities to analyze specific linguistic data. Since our data was already split into training and test components, a pretrained sentiment analyzer VADER (Valance Aware Dictionary and Sentiment Reasoner) was used. As a result, the data is separated into positive, negative, and neutral. These results are stored and later combined with the n-gram part (https://www.nltk.org).

Word Cloud The word cloud is a method to visualize text data analysis. It is based on the repeated number of words or importance of the words. As the importance or the repeated number is increased, the size of the word is increased in the cloud. This visualization method is used to express the general word usage; it is sorted according to the occurrence or importance. Since it looks proper and interesting while show the results, it is used when word determination is important in a project. It appeals to the user with the colors and all the words appear with different sizes. It can be implemented and processed in Python directly using the "matplotlib" and "wordcloud" library (https://www.geeksforgeeks.org/generating-word-cloud-python/). In this chapter, this visualization method has been used in addition to the n-gram's result. The reason for this is to observe the general written words in the dataset.

3.3 *Social Network Analysis*

Social network analysis is used widely in areas such as academic studies, business, education, economy, healthcare, criminology, social fields, etc. With the simplest definition, it can be expressed as "the connection between entities". A network is a concept based on graph theory in mathematics. Networks are made up of nodes and connections between them. Nodes may represent various objects such as people, organizations, units, etc. Links between nodes are called edges which may be directed or undirected [19]. The three basic elements that make up a network are the actors, the relations of the actors with each other, and the structure that emerges from the different combinations of these relations [19].

There are several types of social relationship data in social network analysis. The types of relationships between entities in a social network may be handled in various ways, including directed/undirected, weighted/unweighted-binary, two-part network, temporal data set [20].

Social network analysis can be defined as a set of measures used for the analysis of complex interaction patterns. Individuals, groups, organizations, communities, or countries can be considered as units of analysis. Network analysis enables the comparison and analysis of actors at different units and levels [19]. With social network analysis, questions such as the most influential person in a network, the roles of people in the network, how people participate in the social network, how information is spread in the network, and how people behave in the social environment can be answered. In other words, network analysis may be defined as a set of measures that examine entities that make up the network, the relationships between entities, the relationship models, and interactions between communities. A relationship is defined as "the bond between social entities". It is accepted as one of the basic features that make up a network of relationships. Entities consisting of various node sets at all levels, such as individuals, groups, organizations, societies, cities, countries, computers, organs, can be specified as vertex/edge in network analysis [5].

There are some concepts that are commonly used in social network analysis. Centrality describes the importance of nodes in a network. It is a measure of how a node is connected to other nodes, or in other words, the influence a node has on other nodes. Strategic locations in the network usually have the most important or best-known nodes. Different measures of centrality have been proposed to measure this relative importance. The concept of centrality has various measurement methods, including degree centrality, closeness centrality, betweenness centrality, eigenvector, PageRank. Centrality metrics may identify the most important person or the most central person in the network [5].

Degree Centrality Nodes with more edges according to degree centrality criterion are effective. Degree Centrality is one of the criteria that indicates the extent to which a node is connected to its immediate surroundings and neighbors. Degree centrality is defined as the number of edges connecting a particular node to other nodes. Mathematically, it can be expressed as the sum of each row in the adjacency matrix representing the network. In terms of social networks, those who communicate with more people achieve greater centrality value. Nodes with a high degree of centrality are recognized by other network members as important nodes in a central location in the network [5]. Mathematically, as an example, the degree centrality of a node, say x, can be expressed as:

$$C_D(x) = \frac{\sum_{y=1}^{N}(a_{xy})}{(N-1)}$$

Number of the nodes in the graph is N and the value of axy can be 0 or 1, considering whether an edge is shared between the nodes x and y. Further, axy may reflect the weight of an edge in a weighted graph.

Closeness Centrality According to the closeness centrality criterion, nodes that can spread the information in the shortest time are effective. Closeness centrality is based on the concept of distance. Closeness centrality focuses on how close a node is to other nodes in the network. It measures independence or effectiveness [5]. It can be expressed mathematically as:

$$C(x) = \frac{(N-1)}{\sum_y d(y,x)}$$

N represents the number of nodes and d(y, x) is the length of the path between vertices x and y.

Betweenness Centrality Betweenness centrality criterion, emphasizes that nodes which act as bridges are effective in information transfer. It is based on determining the extent to which a particular node acts as a connector between other nodes in the network. In other words, it highlights nodes that act as a bridge between two or more sets of nodes which cannot communicate with each other. Mathematically, this criterion is calculated by finding the shortest paths between all pairs of nodes in the network, then proportioning how many of these paths pass through a particular node [5].

$$C_B(x) = \sum_{u \neq v \neq x} \left(\frac{\sigma_{uv}(x)}{\sigma_{uv}}\right)$$

The number of shortest paths between vertices u and v is represented in the denominator; and the number of shortest paths between vertices u and v that pass-through vertex x is represented in the numerator.

Eigenvector Centrality According to the eigenvector centrality criterion, nodes with more edges and nodes with a common edge are effective. Eigen vector centrality calculates not only the centrality of a node, but also the centrality of other nodes connected to that node, showing the effective value of strategically connected nodes. It is an expression of the importance and influence of specific nodes in the network. This criterion shows that edges which are considered when calculating the centrality and reflect the relationship of the node with other nodes are not of equally importance. The more centralized nodes it is connected to, the more centralized a node will be. When calculating the eigenvector centrality, the sum of the centrality degrees of the neighbors is considered [5]. For the work described in this chapter, a directed network(Twitter) is considered. The eigenvector centrality measures mainly work for undirected networks. However, knowing this centrality measure provides an opportunity to understand the networks. Mathematically the eigenvector centrality is calculated as follows:

$$C_E(x) = \frac{1}{\lambda} \sum_{y \in M(x)} (C_E(y)) == \frac{1}{\lambda} \sum_{y \in G} (a_{x,y} C_E(y))$$

PageRank One of the most important applications of the Eigenvector approach is the PageRank algorithm. It was produced by Google founders to calculate the essence of webpages from the hyperlink network structure. The method considers each nodes' score, then find their importance [5]. This measure is similar to the eigenvector approach, but it additionally includes the direction of the links between the nodes and their importance. This can be used to identify influencer people. Another difference from the eigenvector measure is that PageRank mainly works for directed networks such as Twitter which is the network studied in this chapter.

To detect trending topics and influence on social media, there are four important aspects to consider, namely, determine top influencers and trending topics, proposing an efficient model for the detection process, collecting tweets, and create a user interface. Each one of them has a significant impact to the whole process and hence the outcome. Determining top influencers and trend topics is the main goal of the work described in this chapter. We set the target that at least the top 5 influencers need to be determined to get proper results. When results are compared with real-life trends and influencers, we want at least 80% of the result matched with the real-life trends and influencers.

Different methods were observed during the literature review. The originality of the work described in this chapter is basically using a distinct method compared to the other works described in the literature. For this reason, we investigated three choices to decide on the method to be employed: combining two different methods and obtaining a unique method, choose more than one method and try each of them and select most efficient one, and finally get a hybrid model from existing data and methods. For this reason, finding a unique and efficient method has effect on the originality of the outcome and results of the work done.

4 Determining Trending Topics and Influential People from Open Source Data

4.1 The N-grams Algorithm with Open-Source Dataset

One of the algorithms employed in this work is n-grams. Details of the n-grams algorithm are explained in the methods section. Initially the open-source dataset is tried on unigram and bigram. Since the words used in the tweets can have different structure, single words, compound words or two words can make sense when put together. So, unigram and bigram results are analyzed jointly. The open-source dataset was obtained from Kaggle. The dataset is about the perspective of retail investors. Sentiment analysis can be applied on this dataset and trending words can be determined. The process produced proper results. The applied algorithm has the following simply steps.

- Extracting and reading the dataset
- Training-test split for the analyzer

- Removing punctuation and stop-words
- Generating n-grams
- Separating words according to sentiments
- Visualize the result

1. Extracting and Reading the Dataset With the help of pandas library in python, the csv/excel/text files can be read directly using the 'read_csv' function. The open-source dataset was saved as csv file to be directly read with the specified function.

2. Training-Test Split for the Analyzer Splitting the dataset into a training and test parts is for sentiment analysis. The ratio of the training-test splits was decided as 60% for training and 40% for test. This is the most common split ratio in machine learning. However, the ratio can be decided by trying various combinations to identify the most optimal ratio.

3. Removing Punctuation and Stop-Words Removing punctuation is a critical point of the analysis. Since most of the sentences can include period, comma, question mark, exclamation mark, quotation mark, etc. Not removing these punctuations from the dataset is expected to affect the results. There is a list of stop words defined by python. It includes words such as of, why, between, and, etc. However, it is possible to use a python function to add extra words to the existing stop words to help removing them from the dataset. For instance, the following statement adds three words to the list: stop_words('english') + ['the', 'food', 'covid']. These stop words are used in almost each sentence and not specify the topic of the sentences. In python, there are libraries for both punctuation and stop words. So, this step can be completed easily. However, in the stop words case, there are few available languages. Turkish is one of the available languages (https://pypi.org/project/stop-words/#available-languages). Since, in this work Turkish tweets will be collected, the available languages are crucial for this part. The other important point is having stop words sensitive for upper and lower case. When additional work are added to the list such as "the", if the sentence includes "The", it won't be remove from the dataset. To overcome this problem without overwhelming the set of words with all possibilities, at the beginning of the algorithm, all the dataset letters were change to the lower case using the 'str. lower ()' function.

4. Generating n-grams There is an algorithm for generating n-grams. In this project, unigrams and bigrams were used with the n-grams defined as 1 and 2, respectively. For unigrams, the algorithm separates sentences to get words one-by-one. For bigrams, the algorithm separates the sentences into combinations of two words each. Since our work concentrates on finding trending topics, unigram and bigram may be considered among the most suitable options.

5. Separating words according to Sentiments With this process, sentences can be separated as positive, negative, and neutral. There is a library in python for this purpose which is defaultdict from collections. The dataset was analyzed using this function to determine three separate lists of positive, negative and neutral. After this

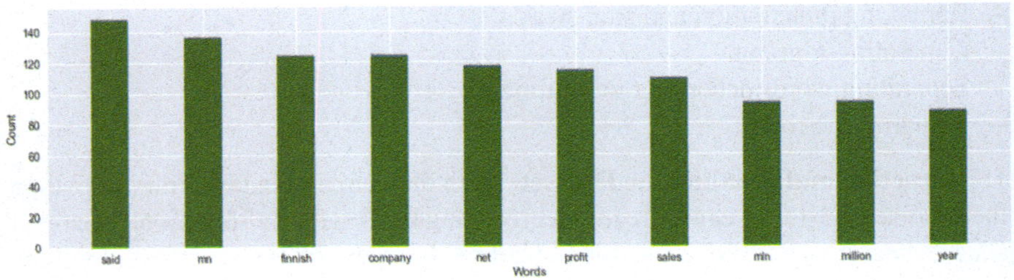

Fig. 2 Top 10 trend words in positive financial news with unigram (ngrams=1)

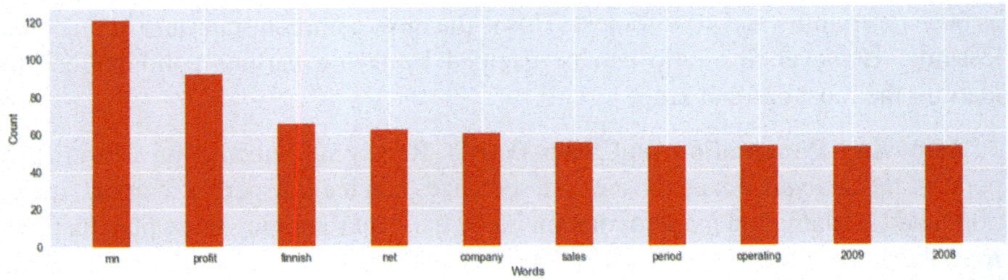

Fig. 3 Top 10 trend words in negative financial news with unigram (ngrams=1)

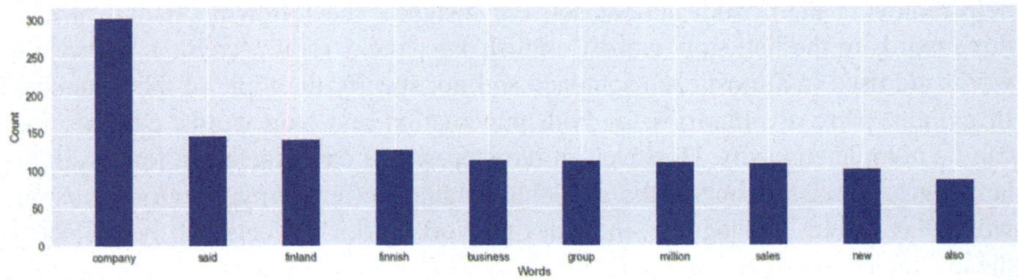

Fig. 4 Top 10 trend words in neural financial news with unigram (ngrams=1)

separation, each list is sorted to find the top 10 most used words. Each of the top 10 words is stored together with its associated number of occurrences.

6. Visualize Result In the visualization part, the matplotlib library was employed to generate bar charts for results. The results were separated into two main parts: unigram results and bigram results. For each of the two types of n-grams, three figures were produced for positive, negative, and neutral. Comparing the unigram (see Figs. 2, 3 and 4) and bigram (see Figs. 5, 6 and 7) results, it can be easily observed that the bigrams showed more reasonable topics, mainly because in the unigram case the words can be apart from the context even though filtering was used. These results demonstrate how the unigrams and bigrams can work properly on sentences. The process did not only determine trending words, at the same time it found and specified if the word was used in positive way, negative way or neutral way.

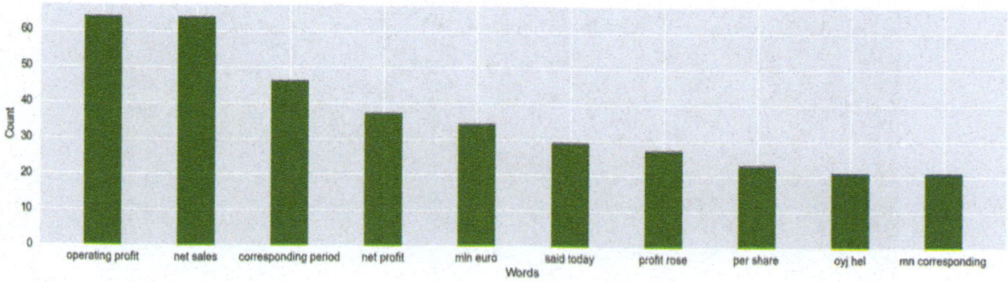

Fig. 5 Top 10 trend words in positive financial news with bigram (ngrams=2)

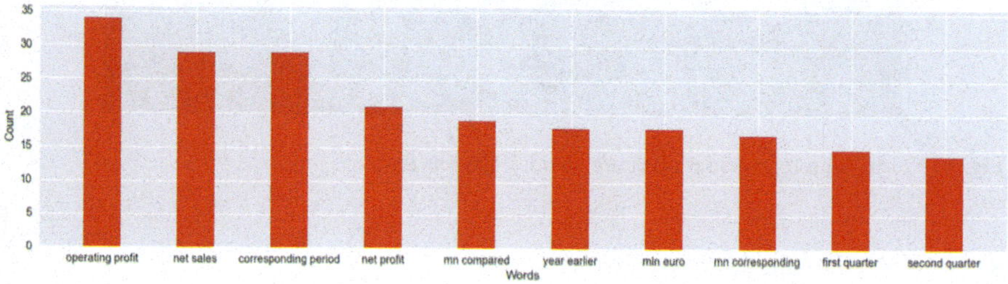

Fig. 6 Top 10 trend words in negative financial news with bigram (ngrams=2)

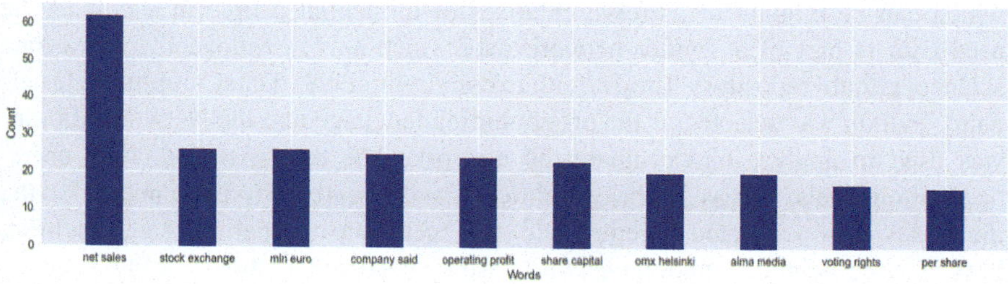

Fig. 7 Top 10 trend words in neutral financial news with bigram (ngrams=2)

4.2 Word Cloud

To get familiar with sentiment analysis, which is relevant for detecting trending topics, the word cloud shown in Fig. 8 was generated using Python by importing the TextBlob and WordCloud libraries. It is basically a visualization method to get most frequent words in a given text. A Covid-19 dataset which includes 1000 tweets from 2020 to 2021 [21]. was used to create a wordcloud. To obtain words that emerged during the pandemic and show the impacts of covid-19 on the social media, some stop words were used such as covid, corona, coronavirus, virus, etc.

From Fig. 8, it can be observed that, 'e businesses have the biggest size in the wordcloud which shows that it is the most used word in the tweets during the pandemic. Also, the words, 'e markets', 'platform', 'e tailers', 'one' draw the attention as the other frequent words in the tweets.

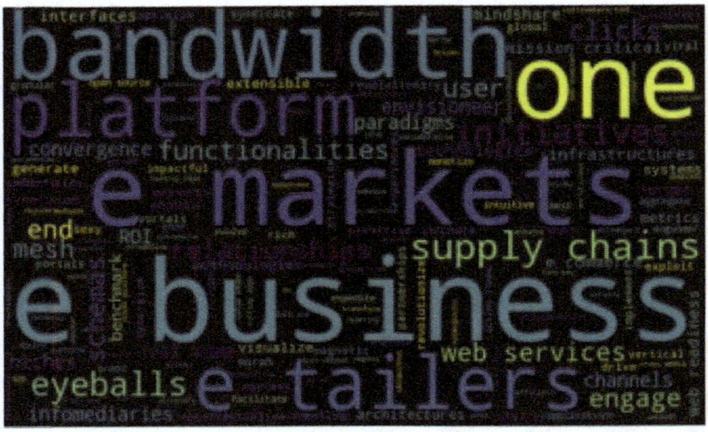

Fig. 8 WordCloud representation using covid-19 tweets dataset

4.3 Centrality Measures

A dataset was used to evaluate centrality measures and detecting important nodes, which can be thought of as representatives of influential people in a graph. The used data is part of a Twitter network data which was downloaded from a data science github repository (https://github.com/trenton3983/DataCamp/tree/master/data). Python was selected as the programming language and the Networkx library was used to analyze and evaluate the network. The data includes some users' occupation levels such as celebrity, politician, and scientist. From the network data, the nodes were extracted together with the metadata occupation, e.g., celebrity, besides, their neighbors and the networks were visualized as shown in Fig. 9.

The centrality measures were obtained from the network and the nodes with the top five scores were recorded from 192 nodes. In order to observe the network from a better perspective, Fig. 10 was visualized as; node colors varies with degree(number of connection/neighbor a node has) and the node size with betweenness, degree, closeness and PageRank measures.

Betweenness Centrality Measure Nodes with the top five values were recorded as 10877, 7331, 24, 36 and 10917. The score of the node 10877 is same with 7331, and the score of the node 36 is same with 10917. This shows that they have the same importance in the network by considering the betweenness measure. However, as can be seen in Fig. 10, the colours of 10877 and 7331 are different which shows they have different degree which means the number of neighbors of them are different. Considering the betweenness centrality for these five nodes, it can be concluded that, the frequency of the nodes in the shortest path among the edges are higher than the others. They can be assumed as important bridges that allow to flow essential information in this dataset. From the graph, it can be observed that these nodes are labelled and have bigger size compared to the others based on the betweenness

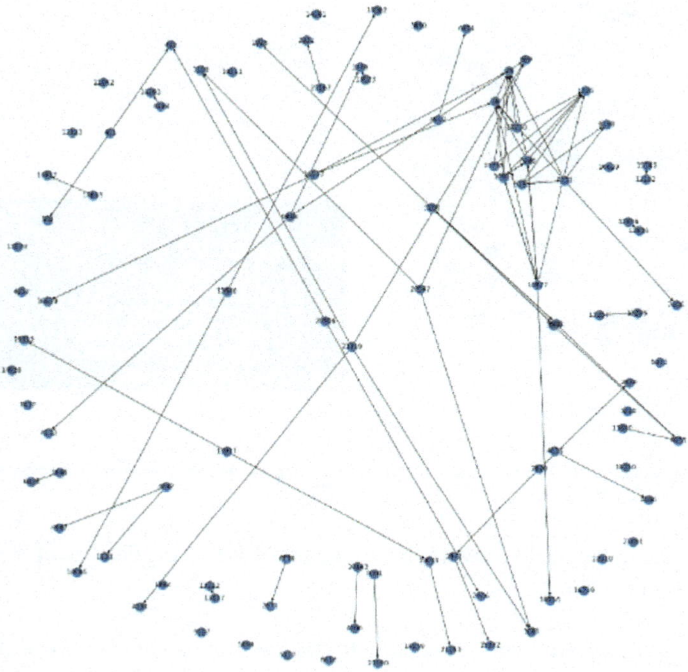

Fig. 9 The default network graph without labels using twitter data, occupation as celebrity

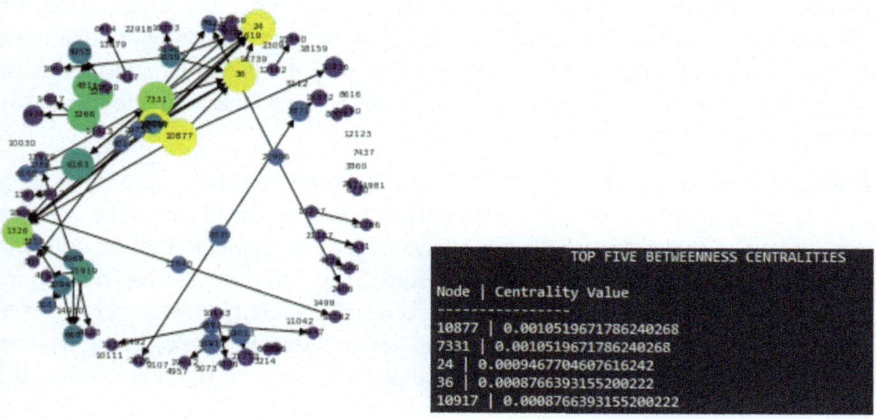

(a) (b)

Fig. 10 Betweenness centrality measure result with twitter data. (**a**) Network (**b**) Centrality measures

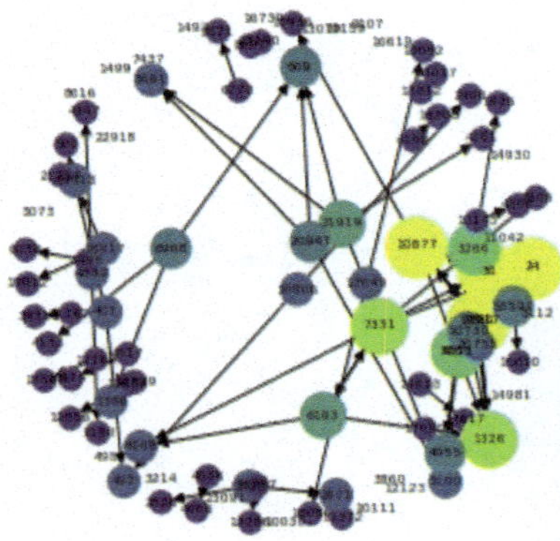

(a)

TOP FIVE DEGREE CENTRALITIES

Node	Centrality Value
24	0.07216494845360824
36	0.07216494845360824
10877	0.07216494845360824
10917	0.07216494845360824
7886	0.07216494845360824

(b)

Fig. 11 Degree centrality measure result with twitter data. (**a**) Network (**b**) Centrality measures

measure. The node 10917 may not be seen clearly due to the density of the nodes in that part of the network.

Degree Centrality Measure Nodes with the top five values were recorded as 24, 36, 10877,10917 and 7886. The score of the top five nodes in Fig. 11 are equal which shows they have the same importance in the network by considering the degree centrality measure. Thus, it can be said that these nodes are the most connected nodes among 192 nodes because the degree centrality measure assumes that the important nodes have many connections and the number of the edges attached to these nodes are higher than the others, In the graph it can be observed that, these nodes are labelled and have larger size compared to the others by considering the degree centrality measure. Some nodes may not be seen clearly due to the density of the nodes in that part of the network.

Closeness Centrality Measure Nodes with the top five values were recorded as 24, 36, 1326, 4955 and 669. From the graph shown in Fig. 12, it can be observed that, these five nodes are labelled and have larger size compared to the others by considering the closeness centrality measure. Thus, they are the top five central nodes among 192 nodes. The measure can be used to identify how quickly an event or a word may spread by the important nodes (influential people) in the network because the important nodes are evaluated as closer ones to the other nodes. It is node 24 in this dataset.

PageRank Nodes with the top five values were recorded as 24, 4955, 1326, 36 and 3265. For PageRank measure, the direction of the links and the importance of those links are considered, then an importance score is assigned to the nodes. Thus, these nodes are the top five important nodes; they are involved in important links.

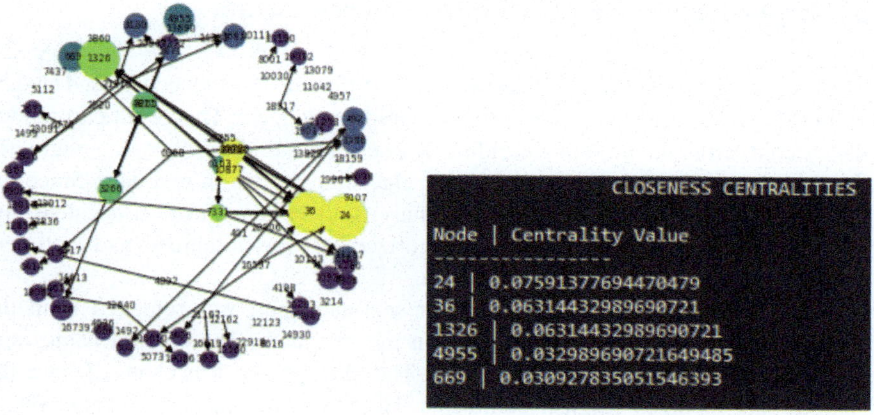

(a) (b)

Fig. 12 Closeness centrality measure result with twitter data. (**a**) Network (**b**) Centrality measures

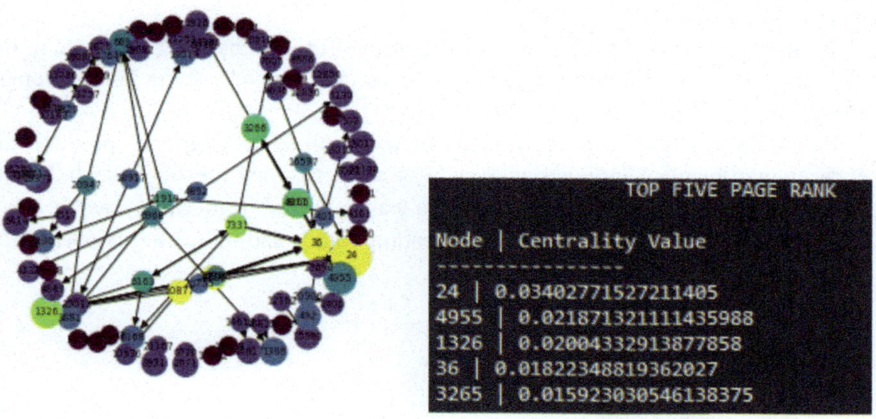

(a) (b)

Fig. 13 PageRank centrality measure result based on twitter data. (**a**) Network (**b**) Centrality measures

The most important node relative to the others is 24 by considering the PageRank measure. As can be seen from Fig. 13, node 36 has the same degree as node 24 (the colour of the node represents the degree), but the importance of node 24 is higher than node 36 which shows that the number of connections or the number of neighbors may not reflect the importance of a node in a graph or influential people in a social network.

5 Experimental Result with the Collected Data

The data was collected from a tweet based on different subject categories. The data included persons' ids, their tweets, reply to information, etc. This result part reports the outcome from the n-gram algorithm. The selected subject is sma_ilac_saglık for this experimental result to observe if this algorithm works on real data. Since the collected data did not include sentiment analysis, there is no positive, negative, and neutral separation. As a next step, all categorized data was combined, and sentiment analysis was applied to the combined data.

The n-gram algorithm was used after sentiment part was removed from the code. The algorithm was edited based on the new data. The column names and the filtering words section were changed. The filtering part was changed since the sample data included English comments. However, the data used in this set of experiments consisted of Turkish tweets. The filtering language was changed from English to Turkish. The unigram and bigram results are shown in Fig. 14 and Fig. 15, respectively. It is possible to also use trigram; this requires only a minor change in the algorithm to produce, but this has been left out because unigram and bigram results were satisfactory.

In this process, we faced a problem regarding Turkish characters because in the data there are symbols rather than some special characters for some Turkish letters like ğ, ç, I, ş, ü, ö.

Sentiment analysis was performed for the sample data used to produce the experimental result. Sentiment analysis was not performed for the Turkish Tweets data used for this set of experiments. Data was collected from Turkey, as the work conducted here focused on identifying trending topics and influential people in the

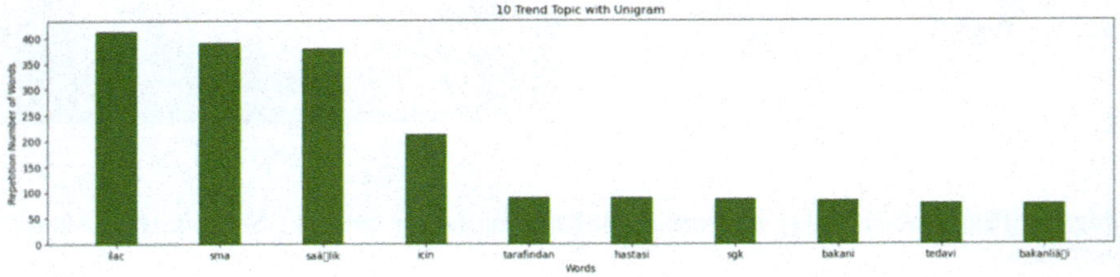

Fig. 14 Unigram result from SMA_Ilac_Saglık tweet dataset

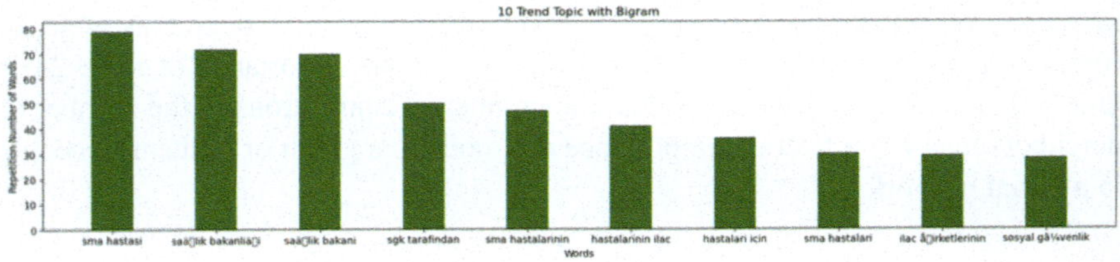

Fig. 15 Bigram result from SMA_Ilac_Saglık tweet dataset

healthcare domain in Turkey. For this reason, the collected tweets contain Turkish characters which are seen as some symbols. This problem was fixed by encryption or other methods. Tweets were divided into separate topics from the collected data. Each group was evaluated separately and then they were all evaluated together. According to this evaluation, an analysis was performed on how to obtain the most efficient result.

5.1 Trend Topics Results

In this project the n-gram algorithm was used to find trending topics. Four types of n-gram, namely, unigram, bigram, trigram and four-gram, were used to produce more detailed and logical data analysis. The n-gram algorithm method is explained with details in the methods section of this chapter. The n-gram algorithm gets the n-gram type as a variable. The four different types of the n-gram results were obtained using a for loop to automate the process.

The collected tweets consist of seven different healthcare related topics. The data includes various features (posting date, tweet text, id, user_id, etc.). In the n-gram, the essential features are tweet text and posting date. Since the data was collected separately based on topic, seven files we produced and then merged as a one combined data file. At the end of this process, there are 6767 tweets to analyze. Since, the data files collected as JSON objects, these files were read in python; then they were converted and saved as CSV files.

Encoding The data included Turkish characters, and this caused some symbol problems. Some of the special Turkish characters are shown in the result as symbols. This problem was solved with special encoding, namely, UTF-8 which is mainly used for electronic communication and known as variable-width character encoding (https://www.utf8-chartable.de/). All csv files were read with this encoding scheme, a solution which worked properly. This way, all the characters appeared in the result without any error.

Stemming After merging, reading, and encoding the data, stemming was done. Stemming is the process of producing morphological variants of a root/base word. Mainly, it returns the word's root form which excludes the suffix, if any. The reason to apply this process is to all similar words considered together while counting the occurrences of trending topics, e.g., 'bakan'-'bakanlar', 'aşı'-'aşının'. The most commonly known stemming libraries generally address English or other common Western languages. Instead we used TurkishStemmer (https://github.com/otuncelli/turkish-stemmer-python) which is a special library for Turkish language stemming. The stemmer library was implemented in n-gram algorithm apart from the program for stemming each word. After stemming, some words were wrongly classified. The reason of this error is that the stemming gets the root of the word and some of the words in Turkish can have different meaning with word suffix and without suffix, e.g., 'aş-aşı', 'bak-bakım' etc. Even after removing the suffix 'bak' word is meaningful, the root of the word 'bakım' which has different meanings based on the context. For this reason, some customization was done. Using a special python function called replace(). In the algorithm firstly we got the words list and applied the mapping process; then replacement with the wrong and original form was done. This customization worked without any error.

Date Selection The data processed in this set of experiments included posting date of the tweet, this additional extension is suitable for the work described in this chapter. The results were presented based on the specified date. A special python function was used to get and specify the date. Firstly, tweet's dates (posted_on column name) were collected with the datetime function. After that, a variable was defined to store the date range. This date range was defined with the date_range function which takes the start date, end date and the frequency of the date range. The frequency can be month, year, week, day even it can be specified as a customized value, such as 3 months or 6 days. Each of the date interval represented a different date range result. For this reason, a for loop was used to obtain the four types of the n-gram result based on each date interval. In this loop, the loc() function was used to compare the posted date and the defined date. The date range was assigned to a variable to be stored as an array; in the for loop this array index was compared with the posted date.

In each of the array indexes, there is one date which has the frequency of a month, e.g., time [2020-01-01, 2020-02-01,2020-03-01,. . .]. For this reason, as depicted in Fig. 16, the date posting in the for loop is compared with the current index and the next index. In each iteration, tweets that have their posting date the same as the specified month were used as a data frame for the n-gram analysis. As a result of this, after all the iterations were completed, multiple results were obtained for each month of each year between 2020 and 2022-03.

Data Visualization Plotly is a graphing library for python. It is open-source, interactive and browser-based library. In this work, plotly was chosen to visualize the result because it is browser-base and interactive. The charts can be managed by the user; it is possible to zoom in and out. There are various properties of plotly charts. This gives an advantage to present the results more user friendly. At the same time, since the results are present on the webpage, it possible to have it easily integrated in the webpage with some extensions.

```
df['created_at'] = pd.to_datetime(df['created_at'])
time = pd.date_range(start="2020-01-01 00:00:00+00:00", end="2020-04-02 00:00:00+00:00", freq="M")
# Filter data between two dates
n = len(time)
for i in range(0, n - 1, 1):
    filtered_df = df.loc[(df['created_at'] >= time[i])
                         & (df['created_at'] <= time[i + 1])]
    temp = df[['full_text', 'created_at']].values
```

Fig. 16 Snapshot of the date specification code

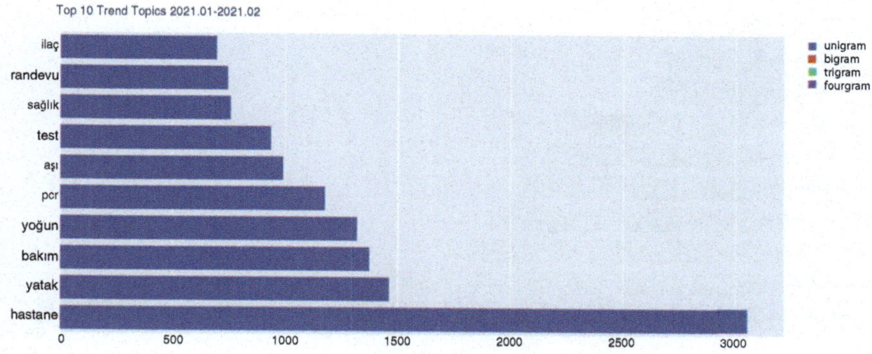

Fig. 17 Trending topics result of unigram (2021.01-2021.02)

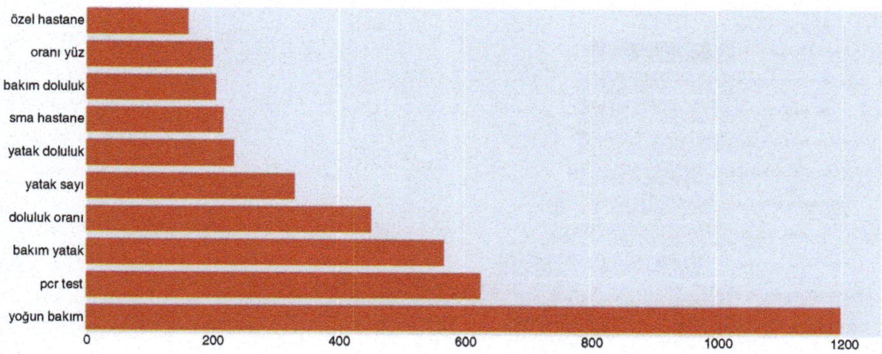

Fig. 18 Trending topics result of bigram (2021.01-2021.02)

Some results of the trending topics are shown in Figs. 17, 18, 19, 20, 21, 22, 23, and 24 for unigram, bigram, trigram and four-gram based on two different dates. However, the real result covered each month between 2020.01 to 2022.03. These have been excluded from this chapter for space limitations to avoid overwhelming the content with more figures than the actual text.

5.2 Social Network Analysis Result

Using the collected tweets data from 01.01.2020 to 22.03.2022 the most influential users were identified for the following six topics related to health: Topic 1 (hastane,aşı,randevu), Topic 2 (hastane,entübe), Topic 3 (hastane,PCR), Topic 4 (kanser,sma,ilaç,sağlık,bakanlık), Topic 5 (yoğun bakımda yer), Topic 6 (yoğun,bakım,yatak). To observe the network from a better perspective, the node

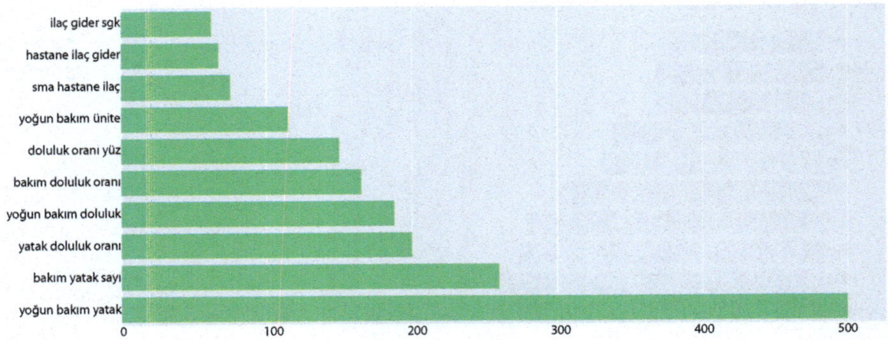

Fig. 19 Trending topics result of trigram (2021.01–2021.02)

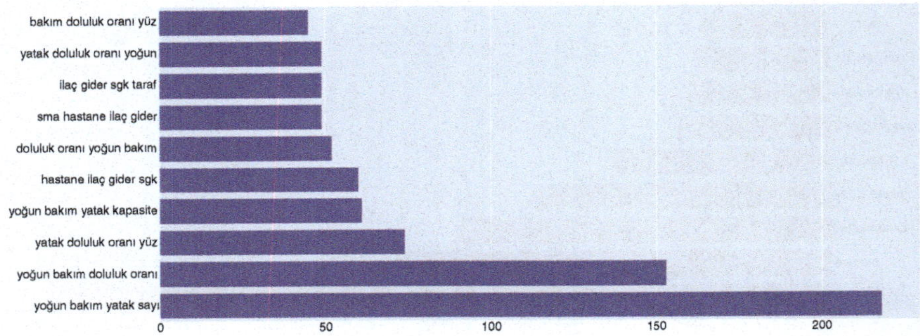

Fig. 20 Trending topics result of fourgram (2021.01–2021.02)

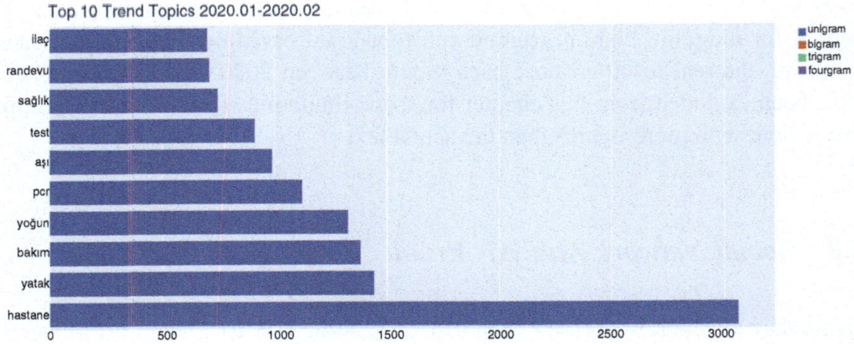

Fig. 21 Trending topics result of unigram (2020.01–2020.02)

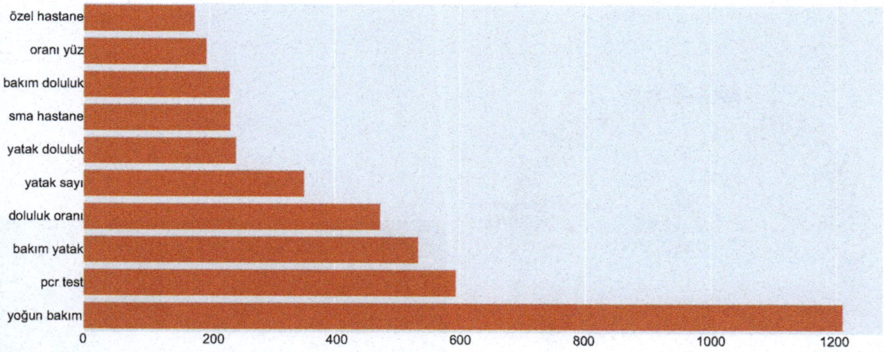

Fig. 22 Trending topics result of bigram (2020.01–2020.02)

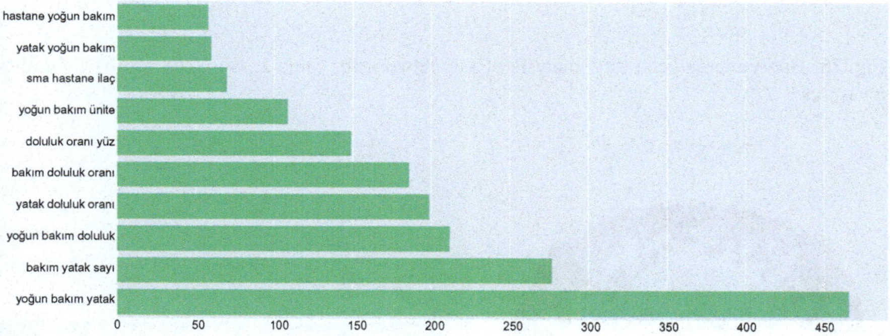

Fig. 23 Trending topics result of trigram (2020.01–2020.02)

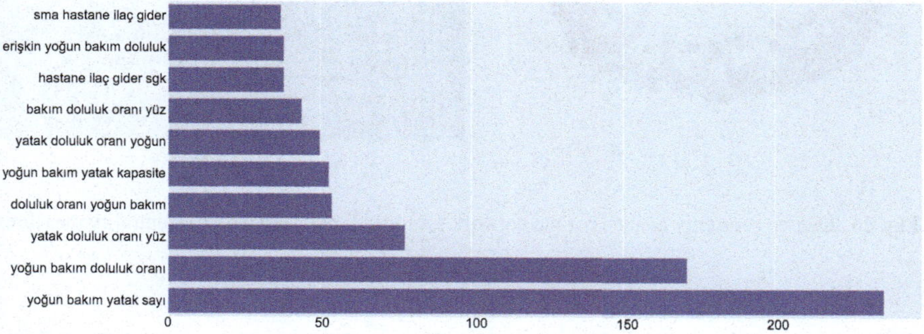

Fig. 24 Trending topics result of fourgram (2020.01–2020.02)

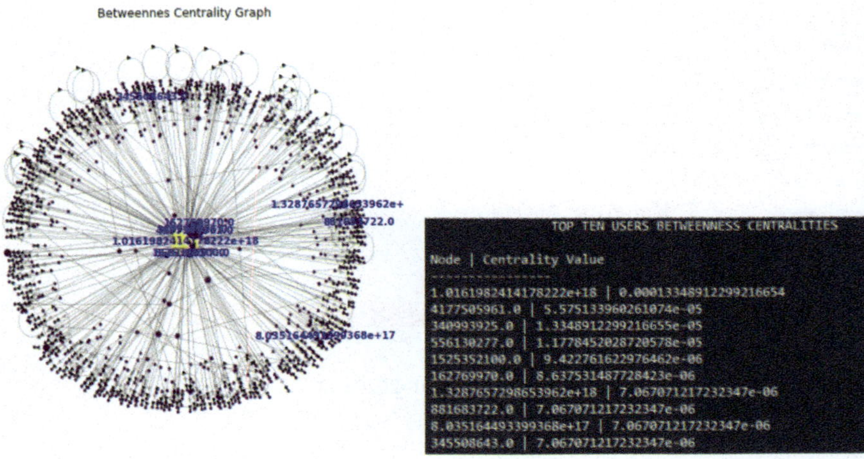

(a) (b)

Fig. 25 Betweenness centrality analysis of the network for topic 1. (**a**) Network (**b**) Centrality measures

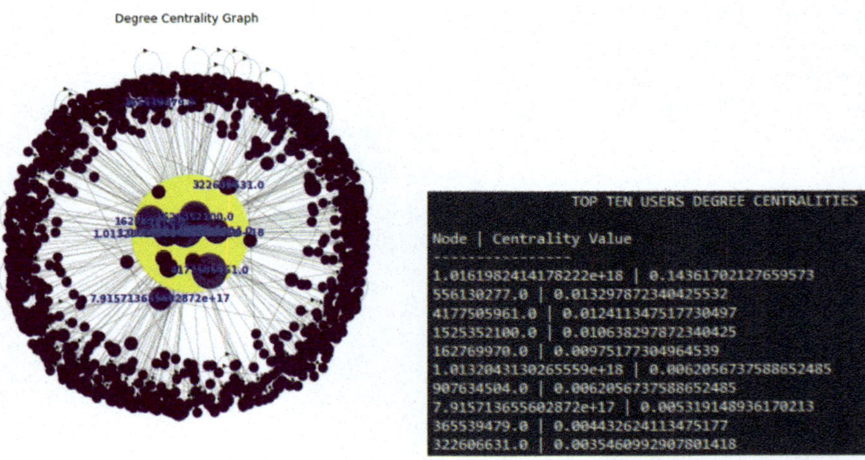

(a) (b)

Fig. 26 Degree centrality analysis of the network for topic 1. (**a**) Network (**b**) Centrality measures

colors varied with degree and the node size with betweenness, degree, closeness, and PageRank measures. In the networks shown in Figs. 25, 26, 27, 28, 29, 30, 31, 32, 33, 34, 35, and 36, it can be observed that the important top ten nodes are labelled and have bigger size compared to the others in all measures. In some of the graphs, the labels of the nodes may not be seen clearly because some of the networks of a topic include more than 2000 nodes, but the importance of the nodes

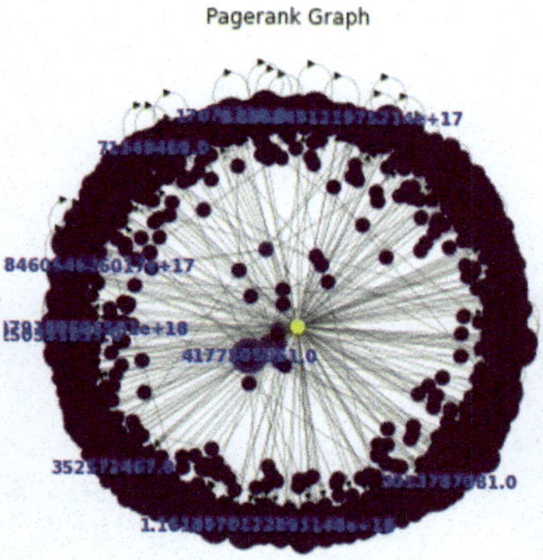

(a)

TOP TEN USERS PAGE RANK

Node	Centrality Value
1.1130079189605581e+18	0.0057779291874436485
71349469.0	0.0057779291874436485
17074738.0	0.0044965392466668665
4177505961.0	0.004310385646829446
7.166846054626017e+17	0.0033860012979936554
250531637.0	0.0032309689347885637
3053787081.0	0.0032309689347885637
9.696849121975214e+17	0.0032309689347885637
352572467.0	0.0032151493058900854
1.1618970122893148e+18	0.0032151493058900854

(b)

Fig. 27 PageRank analysis of the network for topic 1. (**a**) Network (**b**) Centrality measures

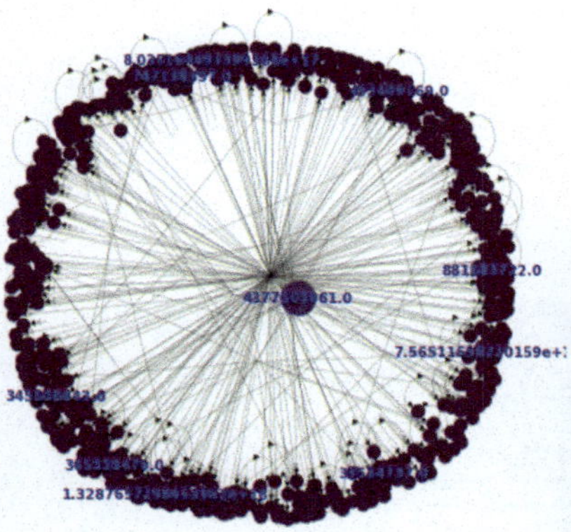

(a)

TOP TEN USERS CLOSENESS CENTRALITIES

Node	Centrality Value
4177505961.0	0.006304176516942474
365539479.0	0.004432624113475178
1.3287657298653962e+18	0.003989361702127659
881683722.0	0.003989361702127659
8.035164493399368e+17	0.003989361702127659
345508643.0	0.003989361702127659
7.56511698220159e+17	0.003989361702127659
33534737.0	0.003989361702127659
493409169.0	0.003989361702127659
747138397.0	0.0035460992907801418

(b)

Fig. 28 Closeness centrality analysis of the network for topic 1. (**a**) Network (**b**) Centrality measures

are also recorded as a table in descending order. While evaluating the users(nodes) in the network, the reply relation between users, reply counts of a user, retweet, favourite, and quote count of the tweets were taken into account.

For topic 1 and topic 2 all the centrality measurements results were added to the report to show that all the measurements were studied, and the most effective one was identified. It was observed that the most reliable measurement is betweenness centrality compared to the others. For example, for the analysis related to topic 1, the same tweets data was used in Fig. 25 (betweenness) and Fig. 27 (PageRank). However, when the nodes were investigated in the real data frame it was found

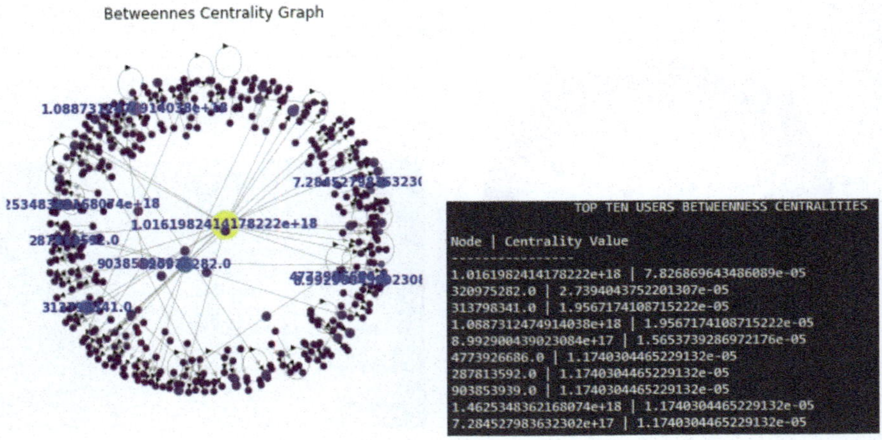

Node	Centrality Value
1.0161982414178222e+18	7.826869643486089e-05
320975282.0	2.7394043752201307e-05
313798341.0	1.9567174108715222e-05
1.0887312474914038e+18	1.9567174108715222e-05
8.992900439023084e+17	1.5653739286972176e-05
4773926686.0	1.1740304465229132e-05
287813592.0	1.1740304465229132e-05
903853939.0	1.1740304465229132e-05
1.4625348362168074e+18	1.1740304465229132e-05
7.284527983632302e+17	1.1740304465229132e-05

(a) (b)

Fig. 29 Betweenness centrality analysis of the network for topic 2. (**a**) Network (**b**) Centrality measures

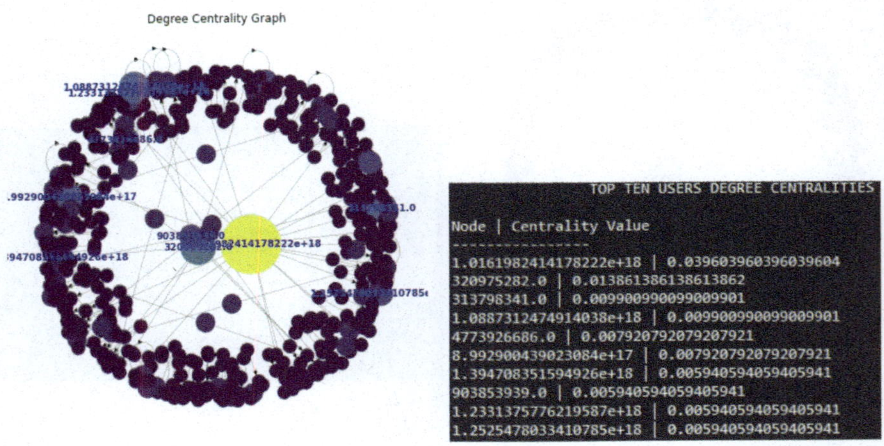

Node	Centrality Value
1.0161982414178222e+18	0.039603960396039604
320975282.0	0.013861386138613862
313798341.0	0.009900990099009901
1.0887312474914038e+18	0.009900990099009901
4773926686.0	0.007920792079207921
8.992900439023084e+17	0.007920792079207921
1.394708351594926e+18	0.005940594059405941
903853939.0	0.005940594059405941
1.2331375776219587e+18	0.005940594059405941
1.2525478033410785e+18	0.005940594059405941

(a) (b)

Fig. 30 Degree centrality analysis of the network for topic 2. (**a**) Network (**b**) Centrality measures

that the connections between nodes and the users' tweets features are negligible for PageRank measures compared to the betweenness measure. For all the topics, the betweenness measure performs the best efficiency compared to the other measures. As mentioned before, this was also investigated in the data frames of all the measures. An illustrating example can be described as follows:

From Fig. 25, it can be observed that the most important node was recorded and marked as an example in Fig. 37. It has more than 600 features, such as retweet

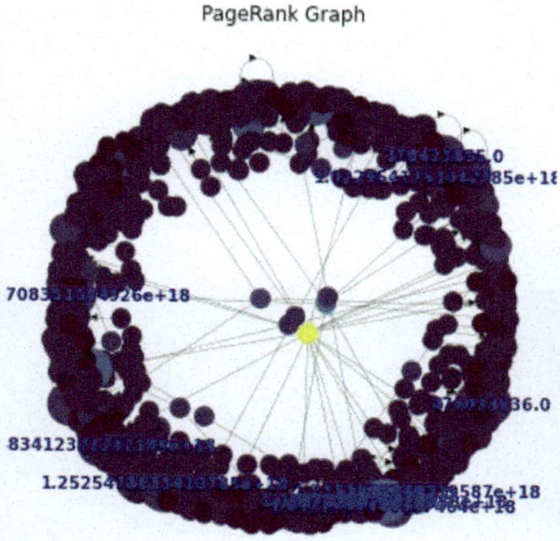

(a)

TOP TEN USERS PAGE RANK

Node	Centrality Value
1.2525478033410785e+18	0.011956621777858821
1.394708351594926e+18	0.007406358440693006
1.2331375776219587e+18	0.006913413245833376
1.0887312474914038e+18	0.006697960115038088
1.0229541751042785e+18	0.006647981217832038
1.483412322232189e+18	0.006647981217832038
1.4331559346337464e+18	0.006647981217832038
536450697.0	0.006647981217832038
476033636.0	0.006647981217832038
178423855.0	0.006647981217832038

(b)

Fig. 31 PageRank analysis of the network for topic 2. (**a**) Network (**b**) Centrality measures

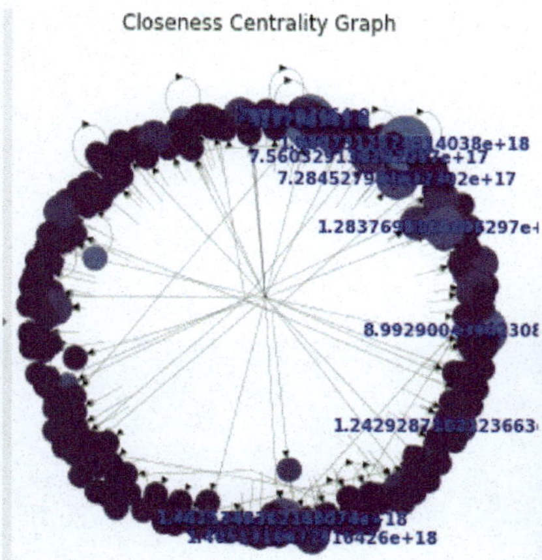

(a)

TOP TEN USERS CLOSENESS CENTRALITIES

Node	Centrality Value
313798341.0	0.009900990099009901
1.0887312474914038e+18	0.009900990099009901
8.992900439023084e+17	0.007920792079207921
1.4625348362168074e+18	0.005940594059405941
7.284527983632302e+17	0.005940594059405941
1.2837698169696297e+18	0.005940594059405941
2999597464.0	0.003960396039603960 4
7.560329136202097e+17	0.0039603960396039604
1.2429287102123663e+18	0.0039603960396039604
1.4051310677916426e+18	0.0039603960396039604

(b)

Fig. 32 Closeness centrality analysis of network for topic 1. (**a**) Network (**b**) Centrality measures

count, favorite count, etc.; this effects user's influence in a network. On the other hand, the most important node in Fig. 27 does not have considerable features count as marked in Fig. 38. The same comparison was done for all the measures, and the results showed that the betweenness centrality measure can be used to find influential users in a network efficiently.

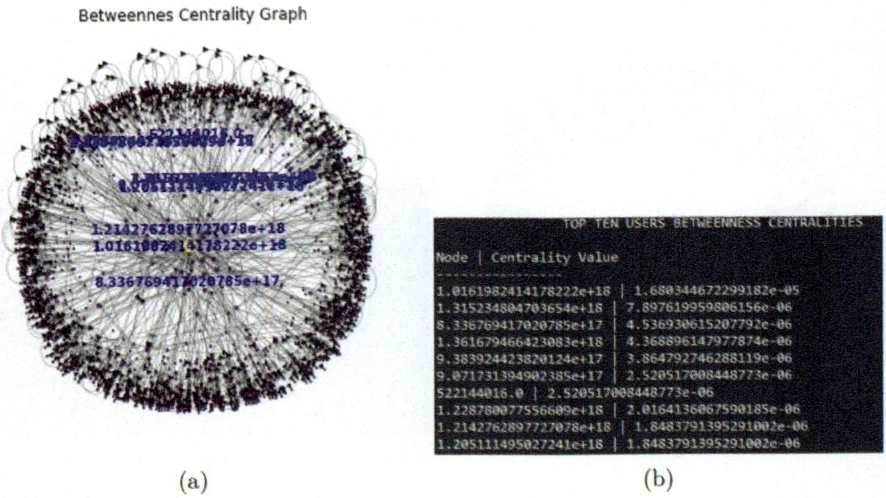

Fig. 33 Betweenness centrality analysis of the network for topic 3. (a) Network (b) Centrality measures

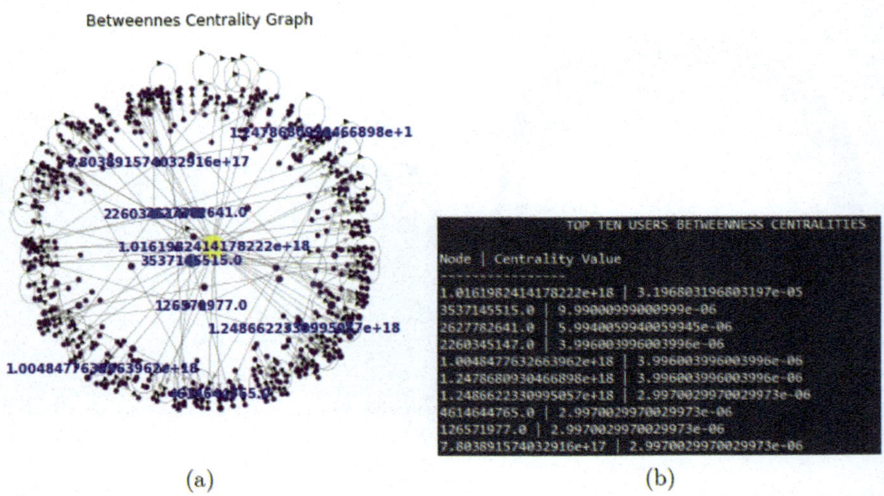

Fig. 34 Betweenness centrality analysis of the network for topic 4. (a) Network (b) Centrality measures

5.3 Web Page

The results are present on the webpage. Webpage is built with flask, python, basic html and css. The flask is used to integrate python code to the webpage. Since both results, namely influencers and trending topics, are obtained from python, the best option to reflect these results to the webpage is integrating python code

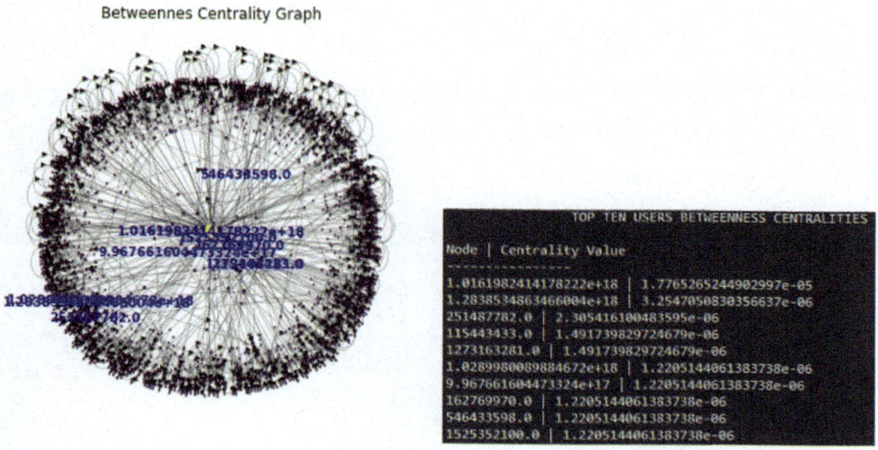

(a) (b)

Fig. 35 Betweenness centrality analysis of the network for topic 5. (**a**) Network (**b**) Centrality measures

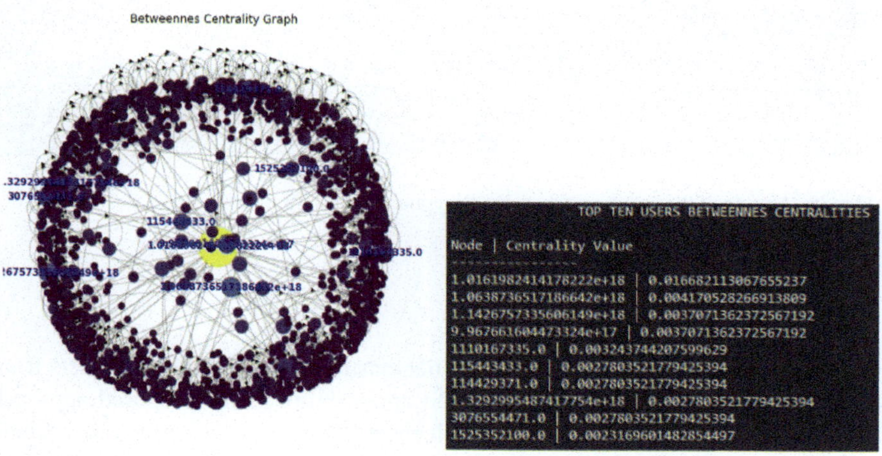

(a) (b)

Fig. 36 Betweenness centrality analysis of the network for topic 6. (**a**) Network (**b**) Centrality measures

to the webpage. The webpage mainly includes three-pages: home, trending topics and influencers. In the home page, there are some details about the project, and some information about project members. In the trending page, there are figures to present trending topics based on dates. In the influencers page, most influencer people are present based on centrality measures.

user id	user id str	geo	oordinate	place	ontributo	uote sta	weet cou	orite cou	ply cour	ote cou
162769970	162769970	nan	nan	None	nan	False	346	7498	139	20
1016198241417822208	1016198241417822208	nan	nan	None	nan	False	681	8621	1401	154
1016198241417822208	1016198241417822208	nan	nan	None	nan	False	655	7621	5470	559
1016198241417822208	1016198241417822208	nan	nan	None	nan	False	656	7885	1325	126
1328765729865396224	1328765729865396224	nan	nan	None	nan	False	0	0	0	0
881683722	881683722	nan	nan	None	nan	False	1	2	0	0
1016198241417822208	1016198241417822208	nan	nan	None	nan	False	1378	12020	1300	177

Fig. 37 A piece of the data frame from the betweenness centrality analysis for topic 1

user id	user id str	geo	oordinate	place	ontributo	uote sta	weet cou	orite cou	ply cour	ote cou
352572467	352572467	nan	nan	None	nan	False	0	0	0	0
1161897012289314821	1161897012289314816	nan	nan	{'id': '5e02a0f0d91c76d2', 'url': 'https://…	nan	False	0	0	1	0
250531637	250531637	nan	nan	None	nan	False	0	0	1	0
250531637	250531637	nan	nan	None	nan	False	1	2	0	0
716684605462601728	716684605462601728	nan	nan	None	nan	False	0	0	0	0
716684605462601728	716684605462601728	nan	nan	None	nan	False	0	0	1	0
3053787081	3053787081	nan	nan	None	nan	False	0	0	0	0
3053787081	3053787081	nan	nan	None	nan	False	0	0	1	0
969684912197521408	969684912197521408	nan	nan	None	nan	False	0	0	1	0
969684912197521408	969684912197521408	nan	nan	None	nan	False	0	0	1	0
969684912197521408	969684912197521408	nan	nan	None	nan	False	0	1	2	0
17074738	17074738	nan	nan	None	nan	False	0	1	0	0
17074738	17074738	nan	nan	None	nan	False	0	15	6	0
1113007918960558080	1113007918960558080	nan	nan	None	nan	False	0	1	2	0
4177505961	4177505961	nan	nan	None	nan	False	0	1	0	0
1113007918960558080	1113007918960558080	nan	nan	None	nan	False	0	0	2	0

Fig. 38 A piece of the data frame from the Pagerank analysis for topic 1

6 Discussion

The results reported in this chapter well demonstrated, using various datasets from different languages, the effectiveness and applicability of the proposed approach for identifying trending topics and influencers. The proposed methods include the n-grams algorithm to identify trending topics and social network analysis to identify influencers. To identify trending topics, n-gram method was used to analyze sentences in tweets as shown in the result part. Sentiment analysis was applied on the experimental result using the first Twitter dataset which was mainly English based. The same was not applied on the Turkish language-based tweets due to the lack of the appropriate tools associated with the Turkish language. This has been left as future work.

The tweets datasets examined in this study were analyzed using unigram, bigram, trigram, and four-gram. Since the words used in the tweets can have different structure such as, single word, compound word or two words, it makes sense when put them together. Thus, four types of the n-gram results were analyzed jointly. The results for the n-grams algorithm showed that, the methods can work proper on sentences. Consequently, n-grams can be helpful while detecting trending topics. A WordCloud was constructed to visualize most common words from the collected tweets data. WordCloud can be used as a colored and attractive representation of frequent words when a user wants to look at trending topics.

To identify influential people on a social network, various centrality measures were applied to the collected Twitter data. The network was analyzed and visualized based on graph theory measures because a user in a social network can be considered as a node and the relationship between the users can be represented by edges in a graph. Important nodes, i.e., influential people, and their centrality values were recorded and visualized in graphs. As can be seen in the results, important nodes are not the same for each measure applied on each of the six topics enumerated in the previous section. When all the measures were compared for all the topics, it was concluded that betweenness is the best choice to analyze and find influential people in a network. For the other measures, when the data frames were investigated, it was found that the features of a tweeter to be an influencer should have significant numbers of trending posts to reach and influence more users in a network. For some of the measures, the inefficient results may be attributed to the lower numbers of influencing tweets. The result may be obtained more accurately with higher number of tweets, this has been left as a future work. Also, more users can be added to the system so that the measures can be evaluated more connectively. Additionally, the tweets which are related to the topics may be increased and more words may be included while gathering them; this way more specific users can be obtained as influencers.

7 Conclusions

The two main objectives of the work described in this chapter, which are determining the trending topics and influential people in the healthcare domain, were achieved. The first part of the work concentrated on identifying trending topics in the healthcare domain using n-grams. The results were obtained from the collected data as unigram, bigram, trigram, and four-gram. The second part of the work described in this chapter was dedicated to utilizing social network analysis to identify influential people the healthcare domain. Results were obtained from the collected data using various centrality measures, including betweenness, degree, closeness, and PageRank. A web interface was designed to present the results obtained and a website was built using flask framework. Using the developed useful tool, the results have been presented and visualized.

As future work, there are several planned activities. First, we plan to apply the same methodology to domains other than healthcare. It may be necessary to adjust and tune up the whole process by considering various aspects of the new domain for the methods to be smoothly applicable. Trend topics and influential people can be determined within topics such as political, economic, education, arts, etc. Second, we may also concentrate on improving the interface design. The user interface that assists in the presentation and visualization of the results was designed and set up as a web page. If it is designed and built as a mobile application, it can be useful for different user profiles. It can also be operated on different platforms and domains.

References

1. Adegboyega, L.O.: Influence of social media on the social behavior of students as viewed by primary school teachers in Kwara State, Nigeria. Mimbar Sekolah Dasar **7**(1), 43–53 (2020). https://doi.org/10.17509/mimbar-sd
2. Yeung, A.W.K., et al.: Implications of Twitter in health-related research: a landscape analysis of the scientific literature. Front. Public Heal. **9**(July), 1–9 (2021). https://doi.org/10.3389/fpubh.2021.654481
3. Albalawi, Y., Sixsmith, J.: Identifying Twitter influencer profiles for health promotion in Saudi Arabia. Health Promot. Int. **32**(3), 456–463 (2017). https://doi.org/10.1093/heapro/dav103
4. Bethu, S., et al.: Data science: identifying influencers in social networks. Period. Eng. Nat. Sci. **6**(1), 215–228 (2018)
5. Salman, C.: Göreceli Kenar Önemi Metodu (A New Network Centrality Measure Relative Edge Importance Method) (2018)
6. Mufid, M.R., Basofi, A., Al Rasyid, M.U.H., Rochimansyah, I.F., Rokhim, A.: Design an MVC model using Python for flask framework development. In: IES 2019 - Int. Electron. Symp. Role Techno-Intelligence Creat. an Open Energy Syst. Towar. Energy Democr. Proc., no. Mvc, pp. 214-219 (2019). https://doi.org/10.1109/ELECSYM.2019.8901656
7. Fiaidhi, J., Mohammed, S., Islam, A.: Towards identifying personalized twitter trending topics using the twitter client RSS feeds. J. Emerg. Technol. Web Intell. **4**(3), 221–226 (2012) https://doi.org/10.4304/jetwi.4.3.221-226
8. Winatmoko, Y.A., Khodra, M.L.: Automatic summarization of tweets in providing Indonesian trending topic explanation. Procedia Technol. **11**, 1027–1033 (2013). https://doi.org/10.1016/j.protcy.2013.12.290
9. Yang, T., Peng, Y.: The importance of trending topics in the gatekeeping of social media news engagement: a natural experiment on Weibo. Commun. Res. (2020). https://doi.org/10.1177/0093650220933729
10. Ora, E.: C.P.M.R.: a methodology for identifying influencers and their products perception on twitter. In: Proceedings of the 20th International Conference on Enterprise Information Systems, Rende (2018)
11. Swit Phuvipadawat, T.M.: Breaking news detection and tracking in twitter. In: 2010 IEEE/WIC/ACM International Conference on Web Intelligence and Intelligent Agent Technology, Washington (2010)
12. Jain, S., Sinha, A.: Identification of influential users on Twitter: a novel weighted correlated influence measure for Covid-19. Chaos Solitons Fractals **139**, 110037 (2020)
13. Aiello, L.M., Petkos, G., Martin, C., Corney, D., Papadopoulos, S., Skraba, R., Göker, A., Kompatsiaris, I., Jaimes, A.: Sensing trending topics in Twitter. IEEE Trans. Multimedia **15**(6), 1268–1282 (2013)
14. Sasa Petrovi, M.O.V.L.: Streaming first story detection with application to Twitter. In: Human Language Technologies: The 2010 Annual Conference of the North American Chapter of the ACL, Los Angeles (2010)
15. Alsolami, A., Mundy, D., Hernandez-Perez, M.: A structured mechanism for identifying political. Influencers on Social Media Platforms: Top 10 Saudi Political Twitter Users World Academy of Science, Engineering and Technology. Int. J. Humanit. Soc. Sci. **15**(4), 366–372 (2021)
16. Oo, M.M., Lwin, M.T.: Detecting influential users in a trending topic community using link analysis approach. Int. J. Intell. Eng. Syst. **13**(6), 178–188 (2020)

17. Hacı, A., Veli, B., İletişim, Ü., Süreli, F., Dergi, E.: Sosyal Medyanın Gündem Belirleyicileri: Twitter'da Gündem Belirleme Süreci Üzerine Bir Sosyal Ağ Analizi Agenda Setters of Social Media: A Social Network Analysis on the Agenda Setting Process of Twitter (2020) [Online]. Available: https://orcid.org/0000-0003-0476-8394
18. Bakan, U., Facebook, Z.E.T., Kelimeler, A.: Sanat Okullarının Twitter Kullanım Karakteristiklerine İlişkin Bir Sosyal Ağ Analizi Perspektifi, vol. 1, pp. 138–155 (2020)
19. Eren, Z., Kıral, E.: Socıal Network Analysıs And Usage In Educatıonal Research. Frames from Educatıon Internatıonal Chapter Book, pp. 308-353 (2019)
20. Gençer, M.: Sosyal Ağ Analizi Yöntemlerine Bir Bakış. Yildiz Soc. Sci. Rev. **3**(2), 19–34 (2017)
21. Gupta, R., Vishwanath, A., Yang, Y.: COVID-19 Twitter dataset with latent topics, sentiments and emotions attributes. arXiv (2020)

The manufacturer's authorised representative in the EU is Springer Nature Customer Service Centre GmbH, Europaplatz 3, 69115 Heidelberg, Germany. If you have any concerns regarding our products, please contact ProductSafety@springernature.com

Printed and bound by CPI Group (UK) Ltd, Croydon, CR0 4YY
28/09/2026
02224085-0018